Writing About Nature

Writing About Nature

A Creative Guide

John M. Murray

University of New Mexico Press
Albuquerque

© 1995 by John A. Murray
Published by arrangement with the author.
University of New Mexico paperback edition 2003
Originally published by Sierra Club Books, ISBN 0-87156-436-X

ISBN-13: 978-0-8263-3085-7

Library of Congress Cataloging-in-Publication Data

Murray, John A., 1954–
Writing about nature / John A. Murray.
p. cm.
Rev. ed. of: The Sierra Club nature writing handbook. ©1995.
Includes bibliographical references and index.
ISBN 0-8263-3085-1 (pbk. : alk. paper)
1. Natural history–Authorship. 2. Natural history literature.
I. Murray, John A., 1954– Sierra Club nature writing handbook. II. Title.
QH14 .M87 2003
808'.066508—dc21

2003007238

1   2   3   4   5   6   7   8   9   10   11   12

Printed and bound in the U.S.A. by Data Reproductions

Design and composition by Adriana Lopez

To Jennifer Felsburg, M.D.

Table of Contents

Preface

Nature is that which we perceive through the senses.

—Alfred North Whitehead

from *The Concept of Nature*

The Tarner Lectures at Trinity College

Cambridge University, 1919

i

Ten years have elapsed since I wrote this book. In that time, the world, both natural and human, has changed in ways that could not have been anticipated in 1993. A host of new national parks and monuments have been formed (most notably, the 1.1. million acre Grand Staircase-Escalante Canyons in southern Utah). Species that had been thriving have now become threatened (such as the Aleutian sea otter). Species that were extirpated have been successfully restored (such as the Yellowstone wolf and the Colorado lynx). Elsewhere around the world, unforeseen events have produced new environmental challenges, as with the social unrest that threatens the last upland gorillas of central Africa, the clear-cutting that is rapidly altering the landscape of southeast Asia, the long-term droughts that are so adversely effecting China, the decline of migratory songbirds in North America and the deep sea fishing that is decimating a sensitive environment that is only partially understood.

Through it all, nature writing has remained a constant, inspiring readers and enriching lives with the grace and beauty of the natural world. A number of stellar works in the genre have appeared over the past decade, including Peter Matthiessen's *Tigers in the Snow*, which documents the decline of the Siberian tiger, E. O. Wilson's *The Future of Life*, which places the events of our time into a larger biological context, and David Quammen's *Song of the Dodo*, which explores island biogeography, both as a scientific fact and as a useful metaphor. These and other books have made it clear that nature writing—as a form of literature, as a school of philosophy and as a mode of political thinking—has become a permanent and essential part of our culture.

As I read back through the fifteen original chapters in this book, while preparing this edition, I noted several areas where the text could be improved by revision, deletion or addition, and have made those changes accordingly. I also saw the need to expand the book further and include a timely discussion of the relationship between nature writing and environmental activism. This new edition, then, provides readers and writers with both an updated and expanded treatment of the subject and one that will, hopefully, continue to be relevant and helpful to them.

ii

Nature is the universe, Whitehead said. Everything in it, from the Yellowstone National Park elk poacher staring at the ceiling of his cell in federal prison to the luminous spiral of the Andromeda galaxy holding at the end of the Hubble space telescope. The word nature derives from the Latin infinitive *nasci*, to be born. That word born is as good a synonym for nature as any. "Nature Writing," however, is a much more recent term, and generally is said to begin with the works of Henry David Thoreau (1819–1862). Most scholars of the field (and "nature writing" is now an academic discipline known as "environmental literature") consider nature writing to include all literary works which take nature as a theme: John Muir's Yosemite essays, Herman Melville's south Pacific novels, John Ford's desert films, Theodore Roosevelt's African travel narrative, Richard Nelson's Arctic Ocean Eskimo studies, Black Elk's autobiography, Blind Lemon Jefferson's blues songs, George Schaller's Mongolian snow leopard studies, Edward Abbey's journals, George Page's PBS nature documentaries, Norman MacLean's elegiac report on the Mann Gulch fire, and many other works from all around the world too numerous to list.

What is the purpose of this book?

To put for the first time under one cover some of what you might like to know about how we nature writers approach the task of creating our works. To give those who can not or will not attend schools the means to learn about the subject on their own, at whatever pace they choose. To provide teachers from high school on up with fifteen lessons to be completed in a semester, written by someone who has taught the subject enough to know what works and what does not work. To share with you what I have learned, am in the process of learning, may never know but can make an intelligent guess about when it comes to the genre.

No single book can hope to accomplish all this, and I will be the first to admit it. But these fifteen chapters at least cut some trail into the subject.

Somewhere along the way I began to dream about a book. A book that had never been written. A book that would have made it easier for me, before I had published anything, to cross that mountain range on one side of which are talented people dreaming of writing and on the other side of which are those who have found a way to write. When David Spinner, my editor at Sierra, and I were chatting on the phone one day in the summer of 1993, the subject turned to handbooks and textbooks, and the ultimate outcome of that conversation is what you are holding in your hands. A book that I wish I had twenty years ago when I began writing those little regional articles on fly-fishing and backpacking for *Outdoor Life*, before I knew anything about writing serious essays and books, but wanted more than anything in the world to learn how to make them. Books are sacred, holy artifacts—I knew that from the time I was a boy and read Thoreau— but the process by which they came into being was always a well-kept secret, shrouded in mysterious fog, and overseen by obscure guild masters and including unknown initiation rituals. No one could provide a compass heading. Think of this book as a field guide. As a set of maps that will help you to explore the country of the imagination, with a bit of explanation on the natural history of the subject along the way. Or, to use a more popular metaphor, think of the book as a user's manual that will explain some of the software you didn't know you had, like the ability to create a metaphor or a simile as powerful as any you've ever admired, or to craft an essay or a story into something permanent and worthwhile, or to submit a book proposal to a publisher and have it seriously considered, and perhaps even accepted.

I hope readers will be moved closer to becoming better writers and publishing their work as a result of using this book (it is a book meant to be kept out and used). Each of you can do it, and in a world of hatred and death and violence, we surely need more essays and books with nature as a theme. The joy of holding a magazine with your essay in it, or a book with your name on it, is comparable to only one experience—holding your new-born child. What you write, if crafted well, will be around after you are gone, like your child, like the earth on which we stand. Knowing that somehow makes life not only more endurable, but also more enjoyable.

As always, I welcome correspondence from my readers at P.O. 102345, Denver, Colorado 80250.

—J. A. M.

Chapter One

The Journal

When [Thoreau began], in October of 1837, to keep a journal, the quarry and substance of much of his best work, we begin to see the whole man as we follow the crowded, highly charged, and rapidly evolving inner life that accompanies the busy outer life and reveals the thoughts behind the eyes of the familiar photographs.

—Robert D. Richardson, Jr.,
Henry David Thoreau: A Life of the Mind

In the autumn of 1988 the editors of *Antaeus*, a noted literary periodical, devoted their issue of some 424 pages to "Journals, Notebooks & Diaries." Among the writers included were Annie Dillard, Gretel Ehrlich, Tess Gallagher, Donald Hall, Rick Bass, Jim Harrison, Ed Hoagland, Joyce Carol Oates, and then-Governor Bill Clinton. The stature and diversity of those featured tells us something. Journals are often utilized by successful writers and thinkers as a means of organizing experience, reflecting on life, and generating material for essays and books. Many nature writers— William Byrd, Henry David Thoreau, John Muir, Aldo Leopold, Edward Abbey, to name just a few—have maintained journals on a regular basis.

Nature writers may rely on journals more consistently than novelists and poets because of the necessity of describing long-term processes of nature, such as seasonal or environmental changes, in great detail, and of carefully recording outdoor excursions for articles and essays. Those American writers who have not been journal writers have often used extended letters in much the same way as journals; Mark Twain and Ernest Hemingway (both accomplished writers about nature in such works, respectively, as *Roughing It* and the *Green Hills of Africa*) are representative of this group. The important thing, it seems to me, is not whether you keep journals, but, rather, whether you have regular mechanisms—extended letters, telephone calls to close friends, visits with confidantes, daily meditation, free-writing exercises—that enable you to comprehensively process events as they occur. But let us focus in this section on journals, which provide one of the most common means of chronicling and interpreting personal history.

The words *journal* and *journey* share an identical root and common history. Both came into the English language as a result of the Norman victory at the Battle of Hastings in 1066. For the next three hundred years French was the chief language of government, religion, and learning in England. The French word *journie*, which meant a day's work or a day's travel, was one of the many words that became incorporated into English at that time. (*Journie* had earlier evolved through the Italian *giornata* from the Latin *diurnata*, which meant a day or the length of a day.) The word journal, originally spelled *jurnal* or *journenal*, sprang into being alongside the older word *journey* and referred to the record of a day's work or a day's travel. By the time of Shakespeare the word journal pretty much had the meaning it has today—a diary of the day's events. Each day of life is a journey in the sense that we travel from the private world of the home to the public arena of work, from the past to the present, from the world of sleep to the world of consciousness, from birth a bit closer to death. The journal offers the writer a moment of rest in that journey, a sort of roadside inn along the highway. Here intellect and imagination are alone with the blank page and composition can proceed with an honesty and informality often precluded in more public forms of expression. As a result, several important benefits can accrue: First, by writing with unscrutinized candor and directness on a particular subject, a person can often find ways to write

2

more effectively on the same theme elsewhere. Second, the journal, as a sort of unflinching mirror, can remind the author of the importance of eliminating self-deception and half-truths in thought and writing. Third, the journal can serve as a brainstorming mechanism to explore new topics, modes of thought or types of writing that otherwise would remain undiscovered or unexamined. Fourth, the journal can provide a means for effecting a catharsis on subjects too personal for publication even among friends and family.

Any discussion of journal writing among nature writers must begin with Henry David Thoreau. It has been estimated that his journals, which span his intellectual life from 1837 (age 20) to his death in 1862 (age 44), contain over one million words. As the epigraph by Robert Richardson indicates, the journals formed an intrinsic component in Thoreau's complicated writing process. From the journals came the earliest drafts of such influential works as "Civil Disobedience," "John Brown's Body" and *Walden*. When the journals were finally published in 1906, they greatly increased Thoreau's stature, not to mention that of the then-fledging discipline of American literature. Thoreau used journals for a variety of purposes, from serious discourse on such subjects as the Mexican War to bits of whimsy such as anecdotes, jokes, and gossip. The vast majority of journal entries are concerned with describing nature, whether it is on the micro-scale of insect life under rocks or on the macro-scale of the dispersal of seeds through the forest over a period of generations. The most studied portion of the journals have been those from the two years (mid 1845–mid 1847) Thoreau spent at Walden while recovering from the tragic death of his brother John to tetanus in January 1842. The Walden sojourn was from the outset a liberating experience, both from grief and from the constraints of "civilized" life, as in this passage selected almost at random from that period (March 26, 1846):

> The change from foul weather to fair, from dark, sluggish hours to serene, elastic ones, is a memorable crisis which all things proclaim. The change from foulnessto serenity is instantaneous. Suddenly an influx of light, though it was late, filled my room. I looked out and saw that the pond was already calm and full of hope as on a summer evening, though

the ice was dissolved but yesterday. There seemed to be some intelligence in the pond which responded to the unseen serenity in a distant horizon. I heard a robin in the distance—the first I had heard this spring—repeating the assurance. The green pitch [pine] suddenly looked brighter and more erect, as if now entirely washed and cleansed by the rain. I knew it would not rain any more. A serene summer-evening sky seemed darkly reflected in the pond, though the clear sky was nowhere visible overhead. It was no longer the end of a season, but the beginning.

One can vividly see in this excerpt how the outer landscape of Walden Pond has become a metaphor for the inner transformations that were occurring in the psyche of the twenty-nine year old naturalist, as the desolate "winter" of grief is replaced by the fertile "spring" of healing.

Journals are important not only for the critical insights they provide scholars but also for the way in which they enlarge our perception and enjoyment of the writer's we love. In their journals writers are seen to be as ordinary human beings like you and me, and this makes them, their writing and their age more accessible. Nathaniel Hawthorne, for example, wrote this lively description of Thoreau in his journal on Labor Day, 1842:

Mr. Thorow dined with us yesterday. He is a singular character—a young man with much of wild original nature still remaining in him; and so far as he is sophisticated, it is in a way and method of his own. He is as ugly as sin, long-nosed, queer-mouthed, and with uncouth and somewhat rustic, although courteous manners, corresponding very well with such an exterior . . . He was educated, I believe, at Cambridge, and formerly kept school in this town [Concord]; but for two or three years back, he has repudiated all regular modes of getting a living, and seems inclined to lead a sort of Indian life among civilized men—an Indian life, I mean, as respect the absence of any systematic effort for a livelihood . . . he seldom walks over a ploughed field without picking up an arrow-point, a spear-head, or other relic . . . as if [the Indian] spirits willed him to be the inheritor of their simple wealth . . .

In the same journal entry Hawthorne describes a rather sad situation later that evening as the impecunious Thoreau begged Hawthorne, a man of relative affluence, to buy his hand-made boat "The Musketaquid" (the same boat on which Thoreau and his brother John had floated the Concord River). By reading Hawthorne's account, and those of Emerson and other members of Thoreau's milieu, we are able to form an image of the writer that both complements and contrasts that found in his own writing.

Similarly, we have Gretel Ehrlich (*The Solace of Open Spaces*), in the issue of *Antaeus* already mentioned, providing us with an unusual look at fellow nature writer Ed Hoagland, who has been called "the Thoreau of our time" and who teaches writing at Bennington College in Vermont. She and Hoagland somehow wind up walking through the annual Greenwich Village Halloween parade in the spring of 1985. Hoagland "is dressed like a fox" and he and Ehrlich "stop midavenue and howl." They are treated to a spectacle of "undulating dragons, twenty-foot-high puppets, [and] dancing skeletons." Each participant in the bacchanal carries a corn stalk as their "pledge of allegiance to maize." While drifting west of Washington Square the two nature writers come upon a "6'8" transvestite dressed in a girl's cheerleading suit [twirling] a baton" and marching to a John Philip Sousa song blasting from a cassette player. Needless to say, this sort of material is not found in the nature essays of either Ehrlich or Hoagland and is refreshing both in its humor and in its verisimilitude. Readers instinctively trust writers who are honest, as we see in the journal excerpts from both Hawthorne and Ehrlich.

Rick Bass, another writer represented in the *Antaeus* issue, published a portion of his journals in a 1991 book entitled *Winter: Notes from Montana*. The book chronicles his new life on a ranch in northwestern Montana from September 13 ("the first overcast day") to March 14 ("I won't be leaving this valley"). Like Thoreau, Bass struggles with how best to lead that doubly enriched "border life" mid-way between the city and the country:

I have to go into town today and shop, do errands, sign papers. If only I could shed that other life, the going-into-town life, like a Cicada, pulling free from a tightening, drying, constraining old shell, a molt. But an old one always seems to grow back. A driving snowstorm, big flakes blowing past, crashing into the woods, swirling in the meadows. They are the currency of winter, and I am the richest man in the world. (January 20)

Above all, the book is a meditation, even a revel, on winter with a purpose and tone somewhat similar to that found in Thoreau's well-known essay "A Winter Walk." The advantage that Bass has in the journal format is that he has no need to artificially compress an expansive subject to fit a literary structure, but can follow the season at his leisure, noting changes as they occur day by day, hour by hour:

I'm from the South, will always be from the South. I'll never get used to snow—how slowly it comes down, how the world seems to slow down, how time slows, how age and sin and everything is buried. I don't mind the cold. The beauty is worth it. It's dark now, and still snowing. An inch, two inches, on the ground, on the car, on the trees, everything. If anything needed doing before the snow came, it should have been done yesterday. It's going to be strange falling asleep tonight, knowing that snow is landing on the roof. It's here. We're here. Nobody's leaving. (November 16)

One of the most prolific journal-keepers of recent memory was Edward Abbey, who began maintaining a personal record in 1946 and made his last entry less than two weeks before his death in 1989. Today you can study these approximately half million words in the Abbey Papers at the Special Collections Library at the University of Arizona, Tucson. They comprise 12.1 linear feet and consist of thirty manuscript boxes (a biographer's dream collection). Colorado nature writer David Petersen (*Racks*, *Among the Aspen*) edited these journals down to about 125,000 words and published them in 1994 as *Confessions of a Barbarian: Pages from the Journals of Edward Abbey*. In an article he published in the scholarly journal *Western American Literature* in May 1993, Petersen described how many of Abbey's most celebrated essays were essentially taken "whole cloth" from the journals, such as the novella-length essay "Down the River" in Abbey's book *Desert Solitaire*. According to Petersen, one of the most striking changes evident in the journals was Abbey's gradual but steady shift from youthful "romantic idealism" to "a pragmatic, defensive, sometimes even cynical carapace." This is somewhat reminiscent of the change scholars have often noticed in several of the key writers of the Romantic period, most notably William Wordsworth, who began his adult life embracing the French Revolution and ended it as a curmudgeonly poet laureate. Petersen refused to edit the journals in order to "improve" or "clean up" Abbey's "image" because

> to do so, to cater to the morality censors who dogged him in life, would be to betray both Ed and his often-stated dedication to candidness in journals. White-washing would also defeat what I perceive as a most important purpose for the publication of *Confessions*—to allow us better to know and understand the intriguing complexities of this man Edward Abbey. A stylistically standardized and politically morally socially philosophically 'correct' Edward Abbey would be— well, no Edward Abbey at all, eh?"

Again, the highest obligation of the writer, or the editor, is to the truth.

Many years ago I took the required graduate seminar in Colonial Literature from Bob Richardson at the University of Denver. Most of the readings were dreadfully boring. Cotton Mather expounding upon man's sinfulness. Michael Wigglesworth philosophizing in his cheerful poem "The Day of Doom." Jonathan Edwards speculating about supernature. Midway through the course Professor Richardson assigned the journals of Samuel Sewall, best known for publishing the first antislavery tract in America, and the "secret diary" of William Byrd, who wrote America's first nature book in 1728 (*History of the Dividing Line*). Only then did the period come to life for the class, and not just because we read with amusement as Samuel Sewall doggedly pursued the widow Katherine Winthrop and William Byrd "rogered" his wife (a euphemism for intercourse). The journals placed the authors in a social, and human, context which was not present in their published writing and which was vital to understanding their larger intellectual contributions to American life. In Byrd's case the private journals were particularly important because they balanced the more formal journals that comprised his *History of the Dividing Line* (in 1728 he was dispatched to settle a border dispute between Virginia and North Carolina and assiduously kept a journal of the trip). Byrd was the first Euroamerican savant to venture into the western wilderness, and return with a description, in journal form, of what he had seen. A new language began to emerge in these wilderness journals—as lean and muscular, as vital and alive, as full of unexpected bounties as the land itself—and it is in this prose that we begin to see the distinctive American voice, the voice of Twain and later of Hemingway, first emerge:

> Not far from our quarters one of them picked up a pair of elk's horns, not very large, and discovered the track of the elk that had shed them. It was rare to find any tokens of those animals so far to the south, because they keep commonly to the northward of thirty-seven degrees, as the buffaloes, for the most part, confine themselves to the southward of that latitude. The elk is full as big as a horse and of the deer kind. The stags only have horns and those exceedingly large and spreading. Their color is something lighter than that of the red deer and their flesh tougher. Their swiftest speed is a large trot, and in that motion they turn their horns back upon their necks and cock their noses aloft in the air. Nature has taught them attitude to save their antlers from being entangled in the thickets, which they always retire to. They are very shy and have the sense of smelling so exquisite that they wind a man at a great distance. For this reason they are seldom seen but when the air is moist, in which case their smell is not so nice. They commonly herd together, and the Indians say if one of

the drove happen by some wound to be disabled fron
ing his escape, the rest will forsake their fears to defen
friend, which they will do with great obstinacy till they are
killed upon the spot. Though, otherwise, they are so alarmed
at the sight of a man that to avoid him they will sometimes
throw themselves down very high precipices into the river.
(October 26, 1728)

Practice Exercises

1. Write a journal that covers one season of the year in the manner of
   Rick Bass's *Winter: Notes from Montana*. Either begin on the equinox
   or solstice and proceed daily through that entire quarter of the year,
   noting the changes that occur as nature moves ineluctably through the
   circle of time.

2. Take a trip and assiduously record all of your impressions in a journal.
   Remember to focus as much on human nature—your inner life and the
   interplay of your companions—as much as on wild nature. Bear in
   mind that every journey inevitably moves toward some sort of climax—a revelation, a decision, a resolution, a turning toward or a turning away. Where is that point in your excursion? Can you anticipate it?
   Or does it occur at an unexpected juncture and surprise you?

3. Write a journal entry in the stream-of-consciousness mode. Do not aim
   so much to write in complete sentences and formal paragraphs as to
   liberate your imagination. Revision can always occur later. Take some
   object or occurence in nature as a point of departure. Write until you
   feel you've exhausted your theme and then write some more past that
   point to see what you can discover. Try to locate some new trails of
   writing and thinking. Push yourself beyond the modes in which you
   normal function.

4. Observe a process of nature and write a journal entry recording your

observations in detail. For example, you might watch the snow falling in a wood, or a storm building over the prairie, or the tide coming in. Search for those small details that will give permanent, universal life to your observations.

5. Write a journal entry that chronicles a powerful childhood memory about nature—your first recollection of nature, your most notable childhood excursion—and then relates it to events and perspectives in your life today. For example, I would write about the time when I was four and my parents found me at the screen door in the middle of the night staring out at the summer lightning bugs. What was I doing there? What does that say about me as a person even at that age? Are there lines of continuity between the child of four and the adult of forty?

6. Attend a conference or a meeting on an environmental theme and write a journal entry—an informal response at the end of the day or program—that records your impressions of the people and their interactions. How are the ways that people interact similar to the ways in which animals that you have seen interact (hierarchal animals like wolves, primates, African lions, some cetaceans)? Do you see lines of connection between human behavior and animal behavior? Are certain sorts of people attracted to certain sorts of issues or animals? Would the same people who attended the Fate of the Grizzly Conference in Boulder, Colorado in 1985 also attend a conference on endangered songbirds? Why or why not?

7. Keep a journal specifically to record your response to various nature books you have read. Which books are successful and which are not? Why? Which authors have something to teach you and which do not? And why? How would you approach the themes differently? Is there a constellation of nature writers—either a regional, stylistic or thematic constellation—to which you believe you belong? Why?

Chapter Two

The Essay

Among my Daily-Papers, which I bestow on the Publick,
there are some which are written with Regularity and Method,
and others that run out into the Wildness of those
Compositions, which go by the Name of Essays.

— Joseph Andrews, "The Spectator"

The word *essay* comes from the French infinitive *essayer,* to attempt. An essay tries to explore a limited subject in a compressed space, the writer acknowledging from the outset that inclusiveness is being sacrificed for ebullience. Traditionally, the French writer Michel de Montaigne has been credited with having invented the essay in his book of short prose pieces entitled, appropriately, *Essais* (1580). Each of these compositions focused on a single topic, such as "the power of the imagination" or "cannibals" or "friendship." However, Montaigne's primacy should be clarified here at the beginning of our chapter, if only to establish a proper historical context. As early as 1597, for example, the British "essayist" Roger Bacon objected to such credit being given to Montaigne and wrote that "Seneca's Epistles [Seneca was a Roman writer of the 1st Century AD] . . . are but Essaies— that is dispersed Meditations . . . Essaies. The word is late, but the thing is auncient." In fact, you could find many other short prose compositions of circumscribed theme in previous ages, whether we look at Caesar, Cicero, or Plutarch, to name just three. This qualification is important because it is fashionable among essayists and students of the essay to consider the form to be Modern—that is, conceived in the Renaissance and brought to fruition in the post-Renaissance period. Nothing could be further from the truth. What is certain, however, is that the nature essay is most assuredly a modern form, and did not exist as a well-defined or regularly practiced genre until the time of Thoreau.

The closest precursors I have found for the nature essay in the century prior to Thoreau are some fascinating prose pieces published in 1750 by Griffith Hughes as *The Natural History of Barbados*. Hughes was the rector of St. Lucy's Parish on Barbados Island in the Caribbean and was also a Fellow of the Royal Society. Subscribers to his book included such luminaries as Samuel Johnson in England and William Fairfax and John Lee in the Virginia Colony. Hughes's book was far ahead of its time in its synthesis of personal essay and natural history writing. Hughes was also the first literary figure to comment directly on the craft of nature essay writing:

> The historical Description of a Country, like its natural Appearance, must needs be attended with Variety. And as, in travelling over it, we must climb high rocky Hills, and pass through dreary Desarts, as well as open Lawns, and flowery Meads; so the Reader must not always expect to be entertained with beautiful Images and a Loftiness of Style. In Variety of Subjects, this must alter with the Nature of the things to be described. . . . This I can with Truth say, that I have not represented one single Fact, which I did not either see myself, or had from Persons of known Veracity.

Hughes definitely sets the tone for future nature essayists by emphasizing in his last sentence the need for complete fidelity to the truth. His meticulous observation and experimentation with a sea anemone—he called it the "animal-flower"—also anticipates the importance of firsthand field study for future writers of the natural history essay. His writing on the anemone, as a synthesis of strong first person narrative and bona fide field science, was nearly a century ahead of its time:

> This surprising Creature . . . hath for a long time been the Object of my own silent Admiration. . . . The Cave that contains this Animal, is near the Bottom of an high rocky cliff facing the se . . . the Descent to it is very steep and dangerous . . . [as] the Waves from below almost incessantly break upon the Cliff, and sometimes reached its highest Summit. As soon as you are freed from this . . . Danger . . . you enter a Cave spacious enough to contain five hundred people . . . From this you enter another Cave, small in Comparison of the former . . . In the Middle of this Basin there is a fit Stone or Rock . . . which is always under water. Round its Sides, at different Depths . . . are seen at all Times of the Year several seemingly fine radiated Flowers of a pale yellow, or a bright Straw-colour slightly tinged with Green: these have in Appearance a circular Border of thick-set Petals, about the Size of, and much resembling, those of a single Garden Marigold . . . I have often attempted to pluck one . . . but could never effect it. For as soon as my Fingers came within two or three Inches of it; it would immediately contract . . . There were strong Appear-ances of Animal Life; yet, as its Shape, and want of local Motion, classed it among Vegetables, I was for some time in Suspense, and imagined it might be an acquatic Sensitive Plant . . . But in what manner the Rays of Light affect these animals, whether by its motion acting upon their whole exceedingly delicate nervous System, which, like the Retina of the human Eye, is in every Part sensitive, is, I believe, inexplicable . . .

This is as fine a piece of personal writing on nature as anything you will find in the writings of Darwin during the voyage of the Beagle in 1832 or in the journals and essays of Thoreau in the 1840s and 1850s. Clearly, Hughes was right on the verge of writing nature essays, and, by certain definitions—the free-flowing questioning of phenomena; the sophisticated use of similes and metaphors ("resembling those of a single Garden Marigold" and "like the Retina of the human Eye"); the strongly present voice of the narrator; the unified, albeit rambling and brief, topical pieces—he probably already was.

It was left to Henry David Thoreau, however, to finally synthesize the diverse sources for the contemporary essay—the private diary, the scientific monograph, the extended letter, the popular magazine article, the conventional essay—into what is now referred to as the nature essay. Thoreau achieved this feat in a series of remarkable early (1842) prose compositions that included "Natural History of Massachusetts," "A Walk to Wachusett," and "A Winter Walk." The latter two essays take a conventional Romantic form—the excursion—and turn it upside down into a new, more comprehensive form that anticipates, indeed that paves the way for, twentieth century excursion essays by writers like Edward Abbey and John McPhee. "Natural History of Massachusetts" is even more daring, as Thoreau forces together all sorts of disparate technical material (the essay began as a review of several new scientific reports) into a prolonged meditation on the essence of nature itself, creating a tradition that later scientifically-trained nature writers like Lewis Thomas and Rachel Carson would inherit. The personal stories included by Thoreau help to make this piece something more than anything European or American literature had seen before.

In his finest essays—such as "Walking," "Wild Apples," and "Autumnal Tints"—Thoreau defined the genre as it now exists today. "Walking," which is perhaps the most influential nature essay ever written, begins with a well-known proclamation:

> I wish to speak a word for Nature, for absolute freedom and wildness, as contrasted with a freedom and culture merely civil—to regard man as an inhabitant, or a part and parcel of Nature, rather than a member of society.

The essay is studded with once-shocking statements now widely quoted and familiar to all lovers of nature:

Life consists of wildness. The most alive is the wildest.

The civilized nations—Greece, Rome, England—have been sustained by the primitive forests which anciently rotted where they stand. They survive as long as the soil is not exhausted.

In literature it is only the wild that attracts us.

In short, all good things are wild and free.

I think that [people who live in the city] deserve some credit for not having all committed suicide long ago.

Eastward I go only by force; but westward I go free.

I feel that with regard to Nature I live a sort of border life, on the confines of a world into which I make occasional and transient forays only.

The West of which I speak is but another name for the Wild; and what I have been preparing to say is, that in Wildness is the preservation of the World.

Give me a wildness whose glance no civilization can endure, — as if we lived on the marrow of [African kudus] devoured raw.

To preserve wild animals implies generally the creation of a forest for them to dwell in or resort to.

Many of these themes came to virtually define the genre of the nature essay—the notion of the American West as a realm of spiritual restoration (as with Theodore Roosevelt's stay in the Badlands following the death of his wife during childbirth); the belief that forests needed to be preserved in order to save wildlife (as with Abraham Lincoln's establishment of Yosemite Valley as a federally designated park); the critique of the crowded, repressed urban life as a kind of death-in-life (as with the jeremiads and polemics of just about every nature essayist of our time); the belief that our destiny is tied up in the wildness (the whole career of John Muir was devoted to this single idea); the concept of the "border life" between the city and the country (with praises still sung to it by writers like Wendell Berry and Sue Hubbel); and the belief that the wild in literature is what animates it (Thoreau's examples suffice: *The Iliad*, the Scriptures and Mythologies, and all the "uncivilized free and wild thinking" in *Hamlet*).

The nature essay in the last decade of the twentieth century is a remarkably diverse and elastic literary form. From the very beginning, writers have been most attracted to the essay because it is so very different from other forms of writing, such as scholarly or journalistic discourse. The essay is natural, open and loose in its structure and various forms. Writers today continue to utilize the form for these reasons. One of the new variations that nature essayists have seized upon is the nonfiction story. This is a self-contained story that, except as it is presented as a largely truthful extended anecdote or excursion, resembles not an essay—a rambling disquisition on a specific theme—but an old-fashioned fictional short story. Rick Bass's first book, for example, is presented by the publisher in the cover notes as a "series of skillful, easy sketches" or "essays" on "the deer pasture," a family deer hunting ranch in Texas. The individual pieces, though, are most often devoted to telling stories about the people of the deer pasture: Cousin Randy, Uncle Jimmy, Werner Schnappauf, and others. I raise this point not to criticize—for *The Deer Pasture* is one of my favorite contemporary nature books—but to let readers know that the boundaries of the essay form in contemporary times are much broader than they have ever been previously. This is a development to be celebrated, for it shows that literature is growing, just as nature grows.

Similarly, the essays of other nature writers, like Terry Tempest Williams, Linda Hasselstrom, and Edward Abbey quite often transcend the traditional form in bold and refreshing ways. Both Hasselstrom and Williams, for example, have used their major essay collections to date (*Land Circle* and *Refuge*, respectively) to focus on the death of a loved one.

Linda Hasselstrom writes in her thirty-four essays (which alternate with thirty-three poems, another innovation) about the death of her husband to Hodgkin's disease, a form of cancer. Terry Tempest Williams chronicles the death of her mother to ovarian cancer, and explores a possible connection between cancer in Mormon women and the open-air atomic testing in Utah during the 1950s. Both use their nature essays to explore nature as a spiritual refuge, as a source of sanctuary and healing in the same sense that Thoreau sought Walden Woods to recover from the death of his brother or, later, Theodore Roosevelt sought the Dakota Badlands following the loss of his wife and mother on the same day. Again, this indicates an expansion of the nature essay form invented by Thoreau in the 1840s into challenging new thematic territory.

Edward Abbey was famous for the daring ways in which he sought new variations on literary forms. In "Gather at the River," for example, Abbey turned the journal of a week-long float down an Alaskan River into an essay, with the daily entries literally forming the structural basis for the nature essay. Another important essay—"Down the River with Henry Thoreau"—took the river excursion motif and radically changed it by introducing Thoreau as a literal companion on the trip, by constantly referring to passages in Thoreau's essays, journals, and letters as if he were a living person speaking these lines in the present tense. In his essay "My Friend Debris" Abbey chronicled a life-time of camaraderie with an eccentric painter from Taos and his female companion, again an innovation on the traditional nature essay. Abbey also expanded several of his book reviews into nature essays, as Thoreau did with "Natural History of Massachusetts." Abbey's efforts included reviews of Paul Horgan's book on Josiah Gregg (of Santa Fe Trail fame) and of *Zen and the Art of Motorcycle Maintenance*. Similarly, he transformed interviews into nature essays (as in his well known interview of Joseph Wood Krutch). In other cases he turned an essay into a sort of jeremiad, as in his polemic on Rene Dubos (who espoused a "pastoral nature" as is found in France and other Mediterranean countries). In another instance, Abbey took the screenplay for a public television special devoted to "Abbey Country" and converted it into a nature essay entitled "TV Show: Out There in the Rocks." Finally, he converted his notes on a trial of nuclear protestors in Colorado into a nature essay that was also a political statement. His willingness to experiment, at times with great vitality, is in the best spirit of the form.

My nature essays have, by comparison, been fairy conventional to date. Every essay has some sort of "hook" to catch the reader's attention in the opening (see Chapter 4); consists of a beginning, middle, and end; includes significant amounts of natural history information; moves toward some discernible climax, which can take the form of a revelation, an image, a rhetorical question, or some other closing device (see Chapter 5); and strives at all times to keep the personal presence of the narrator in the foreground. One of the things I notice most in novice writers is what I call "the disappearing narrator." This occurs when the writer, struggling to incorporate a large amount of scientific information into the text, simply vanishes for a series of paragraphs. Generally speaking, in a personal or nature essay, the narrator cannot disappear for longer than about two paragraphs without the reader becoming uncomfortable. You need that personal "I" as often as possible to remind the reader that this is not scholarly discourse or impersonal journalism, but is, rather, a personal or nature essay. So I always try to process through whatever scientific information I am including several times, at least, in revision, in order to translate it completely from the specialized language of technical discourse into the largely informal idiom of the nature essay. An effective way of facilitating this is to explain the information to a friend, and then to use much the same language, and manner, in the prose of the essay. In this way, you have a more relaxed, conversational approach.

My essays run from ten to twenty pages in length. Not many—except slightly extended book reviews—have been shorter, but a few have gone much longer (the longest nature essay I have ever read is Ed Hoagland's novella-length "Lament the Red Wolf"). My essay on Chickamauga was over thirty pages long. That essay was among the more difficult nature essays I have written. The major problem was that I was attempting to do something that had not been done before (to my knowledge) in a nature essay—I was writing about the relationship between nature and war. The essay grew out of a real life experience—visiting the Chickamauga Civil War battlefield where my great-great-grandfather had been shot and taken prisoner by the Confederates. My task was fourfold: to write about the natural history of the southern Appalachians in northwestern Georgia, to follow my relative's movements over the battlefield on September 19, 1863, to include my four year old son (my companion) in the narrative, and to discuss in a larger sense the nature of war and its effects on humanity and the environment. For both the opening and closing, I searched for something innovative—the opening consists of an extended word-picture of the lush hardwood forests where the soldiers fought, and the closing takes the reader to the gravesite of Martin Luther King in downtown Atlanta, thus providing a look at the larger human rights issues evoked by the trip to the battlefield. Because the essay had so much personal meaning for me, I worked on it, and nothing else, for a solid month. Generally, I prefer to work on several projects at one time, so that I can move to another when I am tired of the one in front of me (see Chapter 3).

Several other essays I've written have sought to expand the form. The first that comes to mind is "The Coral Reef at Akumal," which takes the reader on an open water swim from the beach, out through the lagoon, and into the heart of a coral reef off the Yucatan coast of Mexico; most nature essays on coral reefs begin with a dive from a boat and this one took the reader from the beach, out to sea, and then back again. The second is "Radiant Darkness," which looks at how a natural history habitat group (diorama) of a scene in southern Africa is created. The essay "El Oso Grande" looks at a lifetime of involvement with grizzly bears, from a scene at a museum when I was a boy of nine to my most recent experiences with bears in the Alaska Range. In that essay, as well as in the essay "Wolf Country," which looks at several years of wolf experiences in the Alaska Range, I tried to bring a single species to life through a series of "excursions" into habitat and lifestyle. Both were written from direct experience, which always provides the strongest basis for an essay, or any piece of writing, for that matter.

Samuel Johnson called the essay "a loose sally of the mind" and a "rambling disquisition" in his dictionary, which was the Ur-dictionary for the language. That pretty well says it, even two centuries later. The essay remains one of the most powerful forms available to writers in any language, and, regardless of arguments as to who wrote the first one, the essay will no doubt be around for as long as the human race. The nature essay will always be separate from the journalistic article because it is not nearly so factually-oriented and it is most definitely not impersonal, nor will it ever be confused with the various forms of imaginative writing. It will forever be "wild and free," to use Thoreau's two favorite words in "Walking," a sort of half-tamed country between the rigid world of article-writing and the illusory realm of fiction; the genre in a sense provides a perfect metaphor for the "border life" that Thoreau said existed best mid-way between the city and the country. In fact, literary historians may one day decide that the essay's rise to popularity paralleled and formed a perfect metaphor for the "border life" of suburbs and satellite communities that became increasingly a way of life in America following the Civil War.

Practice Exercises

1. In the old days—and I am speaking here of the sixteenth, seventeenth, and eighteenth centuries—students learned to write primarily by writing out, word for word, entire prose pieces. Sometimes this was done several times in order to imprint the prose rhythms and rhetorical paradigms as deeply as possible. Students then gathered the results of their labors into thick notebooks called "longbooks." The same pedagogical principle is applied in art schools as students study certain museum paintings and replicate them as closely as possible in order to learn craft and technique. Writers such as Christopher Marlowe, William Shakespeare, John Milton, and Thomas Jefferson learned how to more effectively use the English language in this manner. I stumbled upon this technique while preparing anthologies; in order to acquire an active knowledge of the contents, I typed out, word for word, every essay that was in the collection. Reading is, after all, a passive activity. By writing out the essays, I discovered many fascinating new things about them, from how the writers put words together to form sentences to how they welded the paragraphs together to form large structural units. For this practice exercise, select three different essays by writers of varied historical, cultural, or stylistic backgrounds. Either write out or type out these essays, word for word. Upon completing the exercise, write a short essay on what you have learned from the experience. If you find

the technique useful, you may wish to make it a regular habit, as I have. (Suggested essays: N. Scott Momaday's "The Way to Rainy Mountain," Mary Austin's essay "The Land of Little Rain," and Aldo Leopold's essay "Thinking Like a Mountain.")

2. Use a classic essay, such as Henry David Thoreau's "Wild Apples" as a paradigm for an essay of your own. For example, if you were to choose "Wild Apples" as a model, and you lived in Montana, you could write an essay entitled "Wild Blueberries," which would celebrate that wild fruit in the same manner that Thoreau celebrated the wild apples of Massachusetts. Thoreau divided his essay "Wild Apples" into the following components: "The History of the Apple Tree," "The Wild Apple Tree," "The Crab Apple Tree," "How the Apple Tree Grows," "The Fruit, and Its Flavor," "Their Beauty," "The Naming of Them," "The Last Gleaning," and "The Frozen-Thawed Apple." In your essay aim not for slavish imitation of the model, but for emulation and innovation. Try to improve upon the paradigm, and to make the final product uniquely your own. (Other suggested essays: Edward Abbey's "Cape Solitude" from *Abbey's Road* or Sigurd Olson's "Timber Wolves" from *The Singing Wilderness*.)

3. After you have more fully familiarized yourself with the genre of the nature essay—say by reading several anthologies such as Thomas Lyon's *This Incomparable Land*, Frank Bergon's *The Wilderness Reader*, and Robert Finch and John Elder's *The Norton Book of Nature Writing*—write an essay that breaks new ground thematically, structurally, tonally, or stylistically. Try to do something that has never been done before. Head out into the wilderness, so to speak, and scout some new country. Chop some new trail. Take readers into a realm of your personal experience, or a region of nature, that has not been visited by a writer to date. Perhaps you could select something personal as a means of finding uniqueness.

4. Write two brief accounts of a recent nature excursion. Let the first be a factually based journalistic article and the second an impressionistic

essay. When you have completed the exercise, write a third short essay that explains the differences between the two modes of discourse. How do the two approaches formulate different responses in the reader? What are the advantages and disadvantages of each? With which do you as a writer feel more comfortable?

Chapter Three

The Writing Process

On the third finger of my right hand I have a great callus just
from using a pencil for so many hours every day. It has
become a big lump by now and it doesn't ever go away.
Sometimes it is very rough and other times, as today, it is as
shiny as glass...I hold a pencil for about six hours every day.

— John Steinbeck, *The Paris Review*

Every writer has a different approach to the technical aspect of transforming a vividly remembered experience, a flash of inspiration, a vague feeling, or a pile of rain-splashed notes into a polished piece of prose. Thoreau worked meticulously from his journals, indexing them and then performing a cut and paste to create an essay like "Walking" or a chapter in a book like *Walden*. Annie Dillard adopted a somewhat similar process in writing *Pilgrim at Tinker Creek*, as did Barry Lopez in writing his book *Arctic Dreams* and Richard Nelson in writing his book *The Island Within*. Ernest Hemingway was also known for the great care he took in preparing detailed project plans and thoroughly revising every sentence and paragraph (his biographer Jeffrey Meyers once wrote that "It often took Hemingway all morning to write a single perfect paragraph"). Other writers, like Mark Twain, Edward Abbey or Rick Bass, have frequently employed the alternative approach, working quickly with a minimum of revision; hence they often have a more conversational and casual style than those who choose a more formal approach. The choice of how to write, then, is a strategic one with respect to style, and so it is important that a writer choose a method he or she is comfortable with, both in terms of work habits and in terms of literary style.

Thoreau's chief modern biographer Robert Richardson compared Thoreau's writing method to "that of a poet." Richardson elaborated on this interesting statement in detail:

> He started with jottings, perceptions, phrases, short bits often written on the backs of envelopes or other scraps of paper, and often while out walking. Later, back in his room, he would expand the jottings in journal or notebook or, sometimes, letter. Later still, he would work up a lecture or an essay, or return to a familiar subject, pulling together bits, some of which could be quite a few years old. He kept indexes for his notebooks so he could find things in what became an increasingly complicated multivolume writer's storehouse of material. From jottings to journal or notebook, to lecture, to essay was the usual pattern of development, with much of the creative work, the phrase polishing, coming in the journal and notebook stage, as he worked on one passage or image or sentence at a time.

Time and time again in his influential study, Richardson found the seeds to Thoreau's essays in Thoreau's letters or journals. One of the best examples was provided by Thoreau's essay "Walking" (first delivered as a lecture in April, 1851), which could be traced directly to the "Wild Thinking" passage in his journal entry of November 16, 1850. In just five months Thoreau, working with the complicated method described by Richardson above, transformed a single inspired paragraph into the most influential nature essay in American literature, as well as the well-spring of the modern environmental movement.

In the letters and journals you can also find some of Thoreau's most cogent statements on the writing process, two of which illustrate some of the naturalist's favorite thoughts on the matter:

> It is wise to write on many subjects, to try many themes, that so you may find the right and inspiring one. Be greedy of occasions to express your thought. Improve the opportunity to draw analogies. There are innumerable avenues to the perception of the truth . . . Who knows what opportunities he may neglect? It is not in vain that the mind turns aside this way or that: follow its leading; apply it whither it inclines to go. Probe the universe in a myriad points. Be avaricious of your impulses. You must try a thousand themes before you find the right one, as nature makes a thousand acorns to get one oak. (September 4, 1851)

> Write often, write upon a thousand themes . . . Those sentences are good and well discharged which are like so many little resiliencies from the spring floor of our life . . . Take as many bounds in a day as possible. Sentences uttered with your back to the wall. (November 12, 1851)

What Thoreau seems to be saying here is that writing is like a muscle. It must be exercised and kept strong. He advocates constant practice on "many themes," always maintaining the faculty in a state of athletic readiness, always searching for the one topic or point of departure, like that one good acorn, that will yield a work of unity and maturity.

Another writer who shared this passion for "literary exercise" was Mark Twain, who explained his writing method in a famous passage in his autobiography:

> There has never been a time in the past thirty-five years when my literary shipyard hasn't two or more half-finished ships on the ways, neglected and baking in the sun; generally there have been three or four; at present there are five [written

> August 30, 1906. . . . It was [while writing *Tom Sawyer*] that
> I made the great discovery that when the tank runs dry you've
> only to leave it alone and it will fill up again in time, while
> you are asleep—also while you are at work at other things
> and are quite unaware that this unconscious and profitable
> celebration is going on. There was plenty of material now and
> the book went on and finished itself without any trouble.

Many authors, it seems to me, could benefit from taking Twain's advice
and working on more than one project at a time. The writer who focuses
on only one project, often in great frustration, runs the risk of mistaking
the phenomenon described by Twain—the temporary emptying of the
tank—as a total creative block. It is nothing of the sort. Your imagination
is perfectly capable of working on another project; it has only emptied itself
on that one theme. It has many other tanks perfectly full and capable of
being drained.

Twain also lamented the restraints imposed on the imagination by the
conventional writing process itself, i.e., by physically writing a story down
with a pen:

> With a pen in the hand the narrative stream is a canal; it
> moves slowly, smoothly, decorously, sleepily, it has no blem-
> ish except that it is all blemish. It is too literary, too prim, too
> nice; the gait and style and movement are not suited to nar-
> rative. That canal stream is always reflecting; it is its nature,
> it can't help it. Its slick shiny surface is interested in every-
> thing it passes along the banks—cows, foliage, flowers, every-
> thing. And so it wastes a lot of time in reflection.

Toward the end of his life, Twain relied more and more on a secretary, to
whom he dictated his writings in the conviction that this produced more of
the natural, conversational flow. Twain believed that preserving the spontaneity of the spoken word was paramount, in part because the origins of
written literature were in oral narrative. He realized that any piece of prose
that could be spoken effectively would read felicitously on the page.

Edward Abbey, once described the process by which he wrote his best-known work *Desert Solitaire*, which was a memoir of his life as a seasonal park ranger (summer of 1956–summer of 1957) in Arches National Monument (now Park). After years of drifting "from Utah to California to New York to Florida to Nevada, from one marriage to another, from one part-time temporary job to another" Abbey decided that he would finally try to publish an account of his experiences at Arches. At the time (1967) his once promising career as a novelist (*The Brave Cowboy, Fire on the Mountain*) had sort of foundered and he was working as a school bus driver in Death Valley, California. As it turned out, *Desert Solitaire* came almost directly from the extensive journals he had kept at Arches:

> That night [after deciding to write the book] I rummaged through my trunk, dug out my old notebooks and journals, and transcribed by typewriter the entries I had made during those two seamless perfect seasons in the Arches, among the hoodoo rocks and the voodoo silence of the Utah wilderness . . . [I later] mailed the manuscript, Book Rate, to my agent in New York and in January, 1968, on a dark night in the dead of winter, *Desert Solitaire* was published.

This actually is not much of an exaggeration with respect to the writing method employed by Abbey. As was mentioned earlier, the close relationship between Abbey's journals and his essays in *Desert Solitaire* has been well documented by David Petersen, editor of the journals (*Confessions of a Barbarian*). In Abbey's case he worked closely and almost exclusively from his copious "field notes." David Petersen says that "often [Abbey's] essays and articles came from a one-draft typing of those notes, any necessary changes and amplifications being made as he worked. He would then do a pencil-edit of the manuscript and 'mail the whole mess off to become some overpaid editor's problem.'" "Almost never," he said, "did he retype a manuscript, and rarely did he spend more than a day on an article or essay." The speed with which he composed gave Abbey's prose the spontaneity and informality of a personal letter to his readers. He did not edit himself to any great extent—to excise potentially embarrassing phrases or overstatements. He wrote at times almost in the stream-of-consciousness mode, in a highly personal way that presented him with some advantages over those that employ a more structured approach, but also some disadvantages (a tendency to occasional self-parody and incoherence and the impossibility of ever achieving a systematic body of work or thought).

A writer with the opposite approach is Gary Nabhan (*Gathering the Desert*, *The Desert Smells Like Rain*), a Ph.D. in botany from Arizona who works as an ethnobotanist in the Southwest. Nabhan explained his writing method in a forum held on nature writing at the University of Utah in 1988. With him that evening was Ann Zwinger (*Beyond the Aspen Grove*, *Land Above the Trees*), a professional writer whose method was similar to Nabhan's. Nabhan was responding to a question from Ed Lueders, a professor of literature at the University of Utah, with respect to the method by which Nabhan turned his raw material into a polished prose:

> My field notes begin as random observations, not consciously linked by a preconceived theme. At that moment, I don't try to write essay fragments for later polishing. When I am spending a lot of time in the field, it limits my imagination if I record only material related to one theme or interest. But some things intrigue me more than others, and they gather momentum. So must of my field notes aren't done explicitly for essay writing later on; they're just general habit . . . Recently, I broke away from years of botany writing to work on an essay about why domestic turkeys are so dumb. I read everything I could about turkeys, from the 1888 *Standards of Perfection* for poultry breeds to a wonderful paper by Aldo Starker Leopold on the nature of wildness in turkeys. Then I observed turkeys. I went up to Canyon de Chelly and looked at turkey petroglyphs. I looked at a prehistoric turkey effigy . . . I immersed myself in stacks of references on turkey biology, archeology, and domestication. Then, I shoved all the references and notes back into the file drawer and started writing . . . I see what I've retained from my notes and readings that won't go away that I can explain to my mother or

to my five-year-old boy, so that a particular image becomes real to them. After I write the first draft, I pull out all the references and see whether I've correctly interpreted them . . . [A] metaphor that I use is that an essay is an ecosystem—it's not a linear sequence, it has energy flows and nutrients and lives in it that are more than a listing of facts. It is a mosaic of images, and people remember images. I can go from one part of an essay to another and juxtapose two images . . . that creates some kind of connection . . . An essay is obviously linear . . . but you can plant an image in part two that flowers in part four, and introduce its pollinator in part one.

The last image is a particularly valuable one, because it introduces the whole concept of motifs—of stringing certain ideas, some major and others minor, through the narrative, and of connecting them in subtle ways that alert the reader to deeper harmonies. Nabhan's approach is highly organized, but has a number of safeguards built into it—like putting aside the resource material for the initial draft—that enable him to retain the spontaneity so vital to the life of the essay as an art form.

My method of writing—more similar to that of Lopez and Thoreau than of Abbey and Twain—has developed over two decades of trial and error and is fairly straightforward (you will find a detailed case study of this approach in Chapter 16). The first step, after having received an invitation to write for publication or generated an independent writing idea, is to decide whether or not to proceed. This is probably the most important decision, because writing, editing and publication involve an enormous commitment of resources. I then carefully examine the potential idea—whether an essay or book—and make a preliminary assessment as to how much time it will require. The next step, assuming the answer is to go forward, is to establish a production schedule. It is essential (for me at least) to form a writing schedule and rigorously adhere to the self-imposed deadlines in order to get closure on projects; otherwise I would still be working on unfinished essays and books years past when they should have been finished. If you work a long time on a project, you run the risk of becoming bored or frustrated with it, of playing around too much with plot or character in the case of fiction, or of revising it mechanically past the point of organic perfection (and thus lose the spontaneity and the slight imperfections that are essential to any form of life). Many of the best works of literature have been written in relatively limited periods of time; others, like *War and Peace* or *Arctic Dreams* were created over a long period of time, but were written according to a rigorous composition plan that kept the project constantly moving forward into new ground. The final task of the planning stage is to develop an outline that gives the essay or story a beginning, middle and end. Again, I spend considerable time here because it is time well spent—if the outline is successful, I will never work on pieces that will have to be discarded later. Every part completed will fit into the whole, regardless of the order in which the parts are written.

When I begin writing, I always write the opening and the closing first. Many other writers do this (see the epigraph for Chapter 4). These openings and closings remain pretty much the same through any subsequent editing changes. They form, to use a metaphor, the airport from which I depart and the airport at which I will land. The outline then becomes a sort of flight plan from which I ordinarily deviate very little. In this book, for example, I wrote Chapter 1 ("The Journal") and Chapter 16 ("Publication") first (these two chapters took four days, total, to write), and then wrote Chapters 12–15, Chapters 2–7, and Chapters 8–11 in three blocks of time (each block took ten days). The entire book took roughly thirty-two days to write, with most of the production time falling within the four weeks between fall and spring semester during the Christmas inter-term. I would write one chapter of 2,500 to 3,000 words each day, and revise it the next day, as well as gather thoughts for the next chapter on the production schedule. Each day I would work from between four and ten hours on the writing, usually at night (from 7 P.M. to 3 A.M.) when there would be no disturbances (telephone calls and so forth). Some chapters came quickly, because I had my old lecture notes to work from; in other cases I was creating material *ex nihilo* and it required more time. I find it helpful to put all of the finished chapters in one clip or in one file folder beside the computer, so that I can see the book as it develops. This growing stack of pages seems to give me both the pleasure of work completed and the excitement of work to come; this somehow gives me greater incentive to stick religiously to the production schedule and get closure on the project so that I can move on to something else. Like Mark Twain, I also keep three or four projects on the table at one time. In the case of this book, for example, I "switched off" from it several times when I was tired of writing nonfiction and wrote for a day or two on a novel-in-progress and also attended to correspondence and organizational matters for an anthology project.

I believe that just as the complete instructions for building every organism are encoded in the DNA of each of its individual cells, so is a completed work of literature enciphered in every constituent sentence; the key is to find that buried or hidden code and give it complete realization. I have labored long and hard on many an essay or book to try and achieve that elusive feat. Sometimes you come very close and other times you don't come as close as you would have liked, but the important thing is that you brought it along as far as you could. The more years I write, the more I realize the limitations of time and talent, what is achievable and what is not, when I must bring closure on a writing project and get on with something else. That becomes clearer with each passing year and is something for many years I did not understand properly. That realization was liberating in the sense that the writer is finally freed to focus on that part of him which is most productive and worthy of attention.

The writing process is akin to any process of creation—there is pleasure in the conception and much discomfort in the growth period, and then much more work and even a few sleepless nights while the thing is still helpless after the birth. The things we create—these cultural artifacts called essays and books—then go off to become part of the world, to be as alive as trees and deer and rivers and people. The best of them continue to live, in the years after we are gone, and give to our dead names a second breath of life. The very best of them, the truest and most finely crafted, will endure as long as the redwoods or the bristlecones, like the nature epic of *Gilgamesh*, the nature wisdom of *Ecclesiastes*, the "Hymn to the Sun" of the heretic Pharaoh Amenhotep, the nature lyrics of Shih Ching or the quiet beauty of the Upanishads. They become a part of the timeless community of works that never perish. The writing process is a process by which the individual shows allegiance to that larger community of nature and humankind through labor and sacrifice. The writing process involve the offering up a portion of an individual life to become a part of the cultural DNA by which civilization replicates itself, generation after generation. It is a sacred process, akin to the one at the center of every cell, the miracle by which life does not end after a season or a generation, but continues on through the ages.

As you contemplate the writing process, consider some words of inspiration from Thoreau's *Walden*. "Alexander the Great," Thoreau recalled,

> carried the Iliad with him on his expeditions in a precious casket. A written word is the choicest of relics. It is something at once more intimate with us and more universal than any other work of art. It is the work of art nearest to life itself. It may be translated into every language, and not only be read but actually breathed from all human lips;—not be represented on canvas or in marble only, but be carved out of the breath of life itself. The symbol of an ancient man's thought becomes a modern man's speech . . . Books are the treasured wealth of the world and the fit inheritance of generations and nations. Their authors are a natural and irresistible aristocracy in every society, and, more than kings or emperors, exert an influence on mankind . . . How many a man has dated a new era in his life from the reading of a book.

Practice Exercises

1. In writing your next essay, try the method described in the text in which the opening and closing are written first, and the interior sections are then composed. Use a paragraph outline to organize the overall structure before writing. For example, you might take a hike in a local park or forest. The opening will be the beginning of the hike and the closing will be the end of the hike. Write these first, and then provide a narration of the middle part of the experience. If this technique works effectively for you, it might be interesting to apply to other writing projects. If not, you have at least experimented with a different approach that has deepened your understanding of form and craft.

2. Either phone or write a nature writer in your city or region and arrange to interview him or her with respect to the issue of the writing process. You might ask them questions as to whether they use an informal or formal process, whether their approach to the process has changed over the years, and whether they have any particular advise for you on this subject.

3. If it is your habit to work solely on one project at a time, you might want to experiment with the approach described by Mark Twain and work on two projects simultaneously. You may find that this benefits the quality of writing overall on the two; on the other hand you may find your old approach is best for you. The important thing is to experiment, take risks, and acquire as complete a view of your talents and their potential as possible.

4. If you keep a journal, you might wish to experiment with the "cut and paste" method utilized by Thoreau in writing your next essay: index the relevant sections of the journal, gather together those sections, develop transitions, and synthesize them with new material into an organic essay.

5. If it is your habit to work slowly, experiment in your next essay with writing more quickly: establish a production schedule and a deadline and adhere to it as closely as you can. Compute a reasonable number of work hours for each phase of the project—beginning, middle, and end. Fit each specific job into your daily routine wherever space can be found. Proceed accordingly and then evaluate. If, on the other hand, it is your habit to work quickly, try the reverse tack on your next project: work on the material more slowly (you might wish to even set the writing aside for a period of days or weeks and return to it in stages). Again, the point is to examine alternative schemes to writing and see if yours might be improved.

Chapter Four

The Opening

One of the most difficult things is the first paragraph. I have
spent many months on a first paragraph and once I get it, the rest
just comes out very easily. In the first paragraph you solve most of
the problems with your book. The theme is defined, the style,
the tone. At least in my case, the first paragraph is a kind of
sample of what the rest of the book is going to be.
— Gabriel Garcia Marquez, *The Paris Review*

All narratives, all openings, are ultimately grounded in nature, for nature consists of an endless series of stories. We have, for example, the continual pageant of the seasons, the incessant conflicts between predators and prey, the individual battles of life forms to survive, the contests involved in breeding, the cycles of pregnancy in animals and fruition in plants, the epic migrations of whales and wildebeests, the sagas of mass extinction, the passage of the solar days, the building of great storms, the formation and erosion of continents, the waxing and waning of the moon, the transit of the constellations along the zodiac, the coming and going of comets and meteor showers, the life and death of stars and galaxies, the life and death of the universe itself. Each of these stories, which would be stories even if there were no people around to tell them as stories, has an opening, as well as middle and a closing. Here we will focus on the first part, the beginning. When writers say they are suffering from writer's block, quite often what they actually mean is that they cannot find a point of entry into a particular narrative. If they have an opening, ordinarily they will find a way to write the rest of it. But without an opening, they are as stuck as a man who can not find the words to ask a woman out on a date; the romance will not begin until he does. This chapter will look at some common methods to solve that problem. These are techniques as old as rhetoric. They were taught to Julius Caesar and Marcus Aurelius in the Roman schools of rhetoric two thousand years ago, to William Shakespeare and Christopher Marlowe in the grammar schools of Elizabethan England, and to Ralph Waldo Emerson and Henry David Thoreau in the composition classes of pre-Civil War Harvard College, and they are being taught in a lecture hall somewhere in the world even as you read these lines.

An opening paragraph, as Marquez indicated in the chapter epigraph, is important because so much happens in it. The shorter the piece of writing, the more important the opening becomes, if only because the proportions make it so—a 100-word opening paragraph in a 1,000 word sketch bears a greater fractional proportion to the whole than a 100-word opening paragraph in a 100,000 book. I have never spent months on an opening paragraph, as Marquez has, but I have spent weeks. In the old days, when I worked on a portable Royal typewriter, there would be a stack of discarded drafts of a first paragraph beside my typing table, and sometimes that stack would resemble the manuscript to a novella. The opening paragraph for the river otter essay later used as a practice exercise in Chapter 12 went through over sixty drafts. The reason I remember is that I kept all the drafts and counted them when it was over to see how many drafts an opening paragraph requires. Nowadays, using a Macintosh computer, the editing and re-editing of the first paragraph produces no pile of drafts—only a few print outs now and then—but I still spend hours and hours casting and recasting the opening paragraph. So when Marquez observes that the first paragraph "is a kind of sample of what the rest of the book is going to be" and needs to be finely crafted, he is expressing a verity not always sufficiently emphasized. That is why it is so important to prioritize the opening as much as you do the climax and the closing—if any of these three elements is deficient, the whole creation will suffer, perhaps fatally, from the flaw.

Normally the opening accomplishes three things: it establishes authority (by a confident tone and sure-handed command of the style), it makes some sort of a thesis statement (informs the reader as to theme), and it introduces voice (the stance or personality of the narrator). The amount of space required for this to occur varies considerably. Most often in nature essays, the opening is comprised of a unified paragraph of moderate length. In some cases the narrative is set up by a single startling sentence, strong image, one-line joke, powerful rhetorical question, or curious bit of dialogue. In still other cases the opening may reach for a special effect that requires several pages to achieve. An example of the last can be found in one of the best known "nature" novels ever written, Herman Melville's *Moby Dick*:

Call me Ishmael. Some years ago—never mind how long precisely—having little or no money in my purse, and nothing particular to interest me on shore I thought I would sail about a little and see the watery part of the world. It is a way I have of driving off the spleen and regulating the circulation. Whenever I find myself growing grim about the mouth; whenever I find myself involuntarily pausing before coffin warehouses, and bringing up the rear of every funeral I meet; and especially whenever my hypos [nerves] get such an upper hand of me, that it requires a strong moral principle to prevent me from deliberately stepping into the street, and methodically, knocking people's hats off—then, I account it high time to get to sea as soon as I can. This is my substitute for pistol and ball. With a philosophical flourish Cato throws himself upon his sword; I quietly take to the ship. There is nothing surprising in this. If they but knew it, almost all men in their degree, some time or other, cherish very nearly the same feelings towards the ocean with me.

This continues for another five pages, as Melville introduces us to Ishmael, the chief narrator and character in his epic novel. Melville realized that the success of the novel was dependent on Ishmael's credibility. As an experienced novelist, he also knew that he only had about a chapter for Ishmael to win the reader's trust and interest. In the first chapter of about 1,500 words Ishmael introduces himself (a first person narrative) and we find him to be worldly-wise, honest, humorous (joking about "knocking people's hats off"), well-read (reference to Cato), articulate ("a damp drizzly November in my soul"), self-deprecating ("I find myself . . . bringing up the rear of every funeral I meet"), and philosophical ("all men . . . some time or other, cherish very nearly the same feelings towards the ocean"). Above all, we are given a voice that speaks energetically and confidently, that establishes authority and promises a good story (a "baited" opening).

The opposite of the extended opening as seen in *Moby Dick* is the succinct opening, which requires only a handful of sentences or perhaps just one sentence. A writer who is quite adept at the succinct opening is Rick Bass, referred to earlier in the discussion of his book *Winter*. A few examples of his opening paragraphs will illustrate:

> Lucian Hill is a river person; he is a paddling fool. So is his wife of twenty years, Miss Ramona. Lucian's brother, Winfred E., is even worse. ("River People," an essay from the collection *Wild to the Heart*)

> I got this letter, from the Forest Service. They said they would send me out with some of their people, in a forest where grizzlies used to be, and sheep now are, if I'd write about it, and draw the picture. ("The Grizzly Cowboys," an essay from the collection *Wild to the Heart*)

> I am going to tell you about some mountains in northern Utah. ("Magic at Ruth Lake," an essay from the collection *Wild to the Heart*)

There came a big wind the other day, the kind that sweeps through the woods and makes one tree bend way left while the one next to it bows out to the right. It was a wind without order or direction. A wild wind. I could taste the autumn in it. ("Burrisizing," an essay from the collection *Wild to the Heart*)

These are the names of the places I hunt: Buck Hill, the Water Gap, the Burned-Off Hill, Camp Creek, the East Side, the Back Side of Buck Hill, and Turkey Hollow. ("The Deer Pasture," an essay from the book *The Deer Pasture*).

Grandmother Robson makes the best fried chicken in the world. ("The Day Before," an essay from the book *The Deer Pasture*). Ever brush your teeth with Ben-Gay? Wake up with a terrified armadillo in your sleeping bag? No? You've obviously never been deer hunting with Cousin Randy. ("More about Cousin Randy," an essay from the book *The Deer Pasture*) I'm a jeep man. The deer pasture is a jeep place. ("Progeny," an essay from the book *The Deer Pasture*)

What is most striking about these eight opening paragraphs, other than the fact that they are uniformly brief, is that in each case the writer so clearly defines what the topic of the essay will be. Part of the success of Bass's writing—and he is one of the best-selling and most acclaimed nature writers of the younger generation—is that he quickly grabs his reader's interest and trust and then always pulls them along for a worthwhile story or discussion. We learn here that he is going to be writing about a Jeep, about Cousin Randy, about Grandmother Robson, about his favorite hunting places, about autumn, about some big mountains in Utah, about grizzly bears, and about some "river people." There is no uncertainty or ambiguity in the author's mind. Bass has clearly defined the limited terrain to be covered in his essay. He has also used such devices as humor ("Ever brush your teeth with Ben-Gay?"), rhetorical question (same opening), and striking image (tasting the autumn) to "hook" the reader. Bass is a master of the opening, which partly helps to explain the consistent success of his writing in the marketplace and among reviewers, and to underscore the importance of studying his techniques if a short opening will best serve your narrative.

By far, most nature essays begin with a single unified paragraph. Ordinarily, these paragraphs run somewhere between one hundred and three hundred words. In the old days (before Hemingway's revolution of minimalism) opening paragraphs tended toward greater length, and sometimes they still do, but most today fall somewhere between those two extremes. A writer well-known for his effective opening paragraphs was Mark Twain. Twain was, in addition to being our first great humorist and novelist, an accomplished a non-fiction writer about nature with books like *Life on the Mississippi*, *Innocents Abroad*, and *Roughing It*. One of the most brilliant "nature essays," as it were, in *Roughing It* is Chapter 38, which Twain devotes to Mono Lake in the Sierra Nevada Mountains of California. The first paragraph is illustrative of his approach:

> Mono Lake lies in a lifeless, treeless, hideous desert, eight thousand feet above the level of the sea, and is guarded by mountains two thousand feet higher, whose summits are always clothed in clouds. This solemn, silent, sailless sea—this lonely tenant of the loneliest spot on earth—is little graced with the picturesque. It is an unpretending expanse of grayish water, about a hundred miles in circumference, with two islands in its centre, mere upheavals of rent and scorched and blistered lava, snowed over with gray banks and drifts of pumice-stone and ashes, the winding sheet of the dead volcano, whose vast crater the lake has seized upon and occupied.

Like both Melville and Bass, Twain playfully used humor (in this case hyperbole with "lifeless, treeless, hideous desert" and "solemn, silent, sailless sea—this lonely tenant of the loneliest spot on earth") to build a bond a trust and camaraderie with his reader. Twain was also fond of word-picture openings (more on word pictures in Chapter 6), as we see here in his comprehensive description of the complicated landscape around Mono Lake. As with the other examples, there is no doubt in the reader's mind as to what this little nature essay, or chapter, is going to be about: the theme is Mono Lake, "the loneliest spot on earth," a spot that, through the course of the essay, becomes a pretty transparent metaphor for the loneliness and anxiety all travelers experience when they are far from home and in a place that seems desolate and uninviting.

Another approach in the unified paragraph paradigm is the anecdote, the short story or sketch that introduces the subject of the essay. In my essay "El Oso Grande"(Spanish for "The Great Bear"), which chronicles a lifetime of studying grizzlies in literature and in the field, I begin with a story from childhood:

> When I was a boy in Ohio, my classmates and I took yearly field trips to the Cincinnati Museum of Natural History. Toward the end of these tours, we inevitably gathered around a large glass case near the gift shop, and were reluctant to leave that spot for the bright yellow school buses waiting outside in the rain. This exhibit housed the fossilized 13,000-year-old skull of a grizzly bear that had been unearthed in Welsh Cave, just west of the Cumberland Plateau in central Kentucky. According to the descriptive placard, the Midwest had been at that time a boreal woodland and cool taiga parkland not unlike the Interior of present-day Alaska, and had supported a diverse array of now-vanished mammals. The climate then changed, and Ursus arctos horribilis remained only in the major watercourses and mountainous regions of the Far West.

Initially, this essay began with my first view of a wild grizzly while working one summer in college as a wrangler outside Yellowstone National Park. In the second draft, the essay began with a look at the captive grizzly bears at the Denver zoo. Finally I settled on the museum anecdote as the best opening because the story recaptured all the lost wonder of childhood, the innocence that permits us to see things as they truly are, as opposed, very often, to how we have been conditioned to regard them as adults.

The *in media res* (Latin for "in the middle of things") opening is also commonly found in nature essays. This is an opening that begins literally in the middle of the action, dispensing with any preliminary description as to setting, character, or theme. At some point early in the narrative, the writer then fills in the questions that haven arisen in the reader's mind with a flashback or an expository digression. Peter Matthiessen frequently employed these openings in his fast-paced narrative, *The Snow Leopard*, as in this opening paragraph for "October 1":

> The monsoon rains continue all night long, and in the morn-
> ing it is cool and cloudy. Along the trail up the Gandaki, there
> are fewer settlements, fewer stone huts in which travelers may
> take shelter, and with the north wind comes the uneasy feel-
> ing that, in this autumn season, we are bound into the wind,
> against the weather. Down the river comes a common sand-
> piper, the Eurasian kin of the spotted sandpiper of home: it
> teeters and flits from boulder to black boulder, bound for
> warm mud margins to the south. I have seen this jaunty bird
> in many places, from Galway to New Guinea, and am
> cheered a little when I meet it again here.

The *in media res* beginning creates a degree of suspense in the reader's mind, in that he and she is pulled into the action immediately, and then is set to wondering as to setting, character relationships and possible out-comes. Here Matthiessen, after leaving us the night before with a short geological history of the Himalayas, dispenses with any description of breaking camp and immediately puts us on the trail, moving toward Gandaki, observing the bird life, making new discoveries at every bend of the canyon.

The best openings act on the reader's mind in the same way that the fly-fisherman's imitation mosquito operates on the brook trout hunting for fallen mosquitoes in the shadowy backwaters of the beaver pond. They create the illusion of reality so effectively that the reader is pulled into the dream, into the artificial world being created with words. We are hooked into the narrative, drawn closer and closer to the climax, whatever and wherever that might be. To change the metaphor, the best openings convey to us the excitement that attends the beginning of any journey, for an essay or a book is most definitely a trip from one place on the landscape of the mind to another. We read that first paragraph, or that first page, or that first chapter, and we cannot put the book down, as in this opening paragraph, from Edward Abbey's *Desert Solitaire*:

> This is the most beautiful place on earth. There are many such places. Every man, every woman, carries in heart and mind the image of the ideal place, the right place, the one true home, known or unknown, actual or visionary. A houseboat in Kashmir, a view down Atlantic Avenue in Brooklyn, a gray gothic farmhouse two stories high at the end of a red dog road in the Allegheny Mountains, a cabin on the shore of a blue lake in spruce and fir country, a greasy alley near the Hoboken waterfront, or even, possibly, for those of a less demanding sensibility, the world to be seen from a comfortable apartment high in the tender, velvety smog of Manhattan, Chicago, Paris, Tokyo, Rio or Rome—there's no limit to the human capacity for the homing sentiment. Theologians, sky pilots, astronauts have even felt the appeal of home calling to them from above, in the cold black outback of interstellar space.

Practice Exercises

1. Study each of the essay openings in a classic work of nature writing. Which, if any, of these paradigms (unified rhetorical, succinct, anecdote, *in media res*, rhetorical question, striking image, metaphor or idea, humor, extended, and so on) are applicable to essays you are now either contemplating or working on? Books you might consider reading include Annie Dillard's *Pilgrim at Tinker Creek*, Joseph Wood Krutch's *The Desert Year*, Ann Zwinger's *Beyond the Aspen Grove*, Henry Beston's *The Outermost House*, Roger Caras's *Monarch of Deadman Bay*.

2. Take a well-known nature essay, block out the paragraph the author wrote for it, and compose your own opening. Use a fresh, original approach as you aim for a different point of entry than that chosen by the author. What are the advantages and disadvantages of yours and the original? What have you learned about the importance of openings from the exercise? Essays that might work well for this exercise include Edward Abbey's "The Reef" from *Abbey's Road*, Rachel Carson's "Wind and Water" from *The Sea Around Us*, or John McPhee's "The Encircled River" from *Coming Into the Country*.

3. Do your openings tend toward a dreary sameness? Do you find yourself using the same paradigms over and over again? Experiment with a different type of opening for one of your completed or published essays. You might, for example, employ a baited opener (one in which you hint at or promise to discuss something of interest later in the narrative), a funnel opener (in which you begin with a general statement and gradually focus on the specific theme of the essay), or one of the other approaches discussed in the chapter.

4. Like all writers, nature essayists employ symbols in their work, as when Peter Matthiessen uses the snow leopard in his work by the same name to symbolize his dead wife, or Edward Abbey uses the desert in *Desert Solitaire* to symbolize the desert of his early middle years, or when Terry Tempest Williams in *Refuge* uses a dying salt marsh to symbolize a dying province of the soul. Frequently, these central symbols are introduced in some striking way in the opening paragraph or paragraph block. Write an essay in which you do just that, reveal the unifying symbol of the essay in some subtle or significant way in the opening.

5. One of the most effective openings is formed by asking a rhetorical question, such as "Why would any rational person participate in an annual prairie dog kill?" or "How many dead grizzly bears will it take before the park service recognizes we have a problem?" or "Where are all the Pacific salmon going?" These openings work well because

they immediately establish a relationship with the reader, who has been asked a question and naturally formulates a response. Write an opening paragraph or paragraph block for an essay on a controversial subject that is based on the rhetorical question paradigm. Sometimes a surprising or stirring question is particularly engaging.

Chapter Five

The Closing

"A bad beginning makes a bad ending."
— Euripedes, *Aeolus*

Like the opening, the closing is ultimately based on the great cycles and stories of nature: the end of a season, the conclusion of a solar day or a night,
the close of a lunar cycle, the last act in a species' battle for survival, the
final moments of a long migration, the expiration of a blizzard or hurri-
cane, the sudden bolt of lightning that starts the ancient forest burning, the
slow transformation of a volcanic island into a coral atoll, the long sput-
tering finale of a meteor across the night sky, the explosive death of a star
or a galaxy, the ultimate fate of the cosmos itself. These closings exist, have
existed and will exist, separate from the human experience. They are part
of the elemental narrative structure of the universe, which is based in turn
on linear time and causality. Whereas the opening evokes images and asso-
ciations of birth and the awakening of consciousness, the closing summons
up death and the extinction of consciousness. Readers are very sensitive to
the closing of a narrative, and, regardless of the strength of the rest of the
piece, will feel considerable disappointment if the writer fails them at this
crucial moment. It is probably not an exaggeration to say that if the closing is particularly ineffective, the essay or story will not endure. It is not

enough to simply tell a story, and then make some sort of a hasty or awkward exit around the corner. The closing must be natural and graceful and have a movement that emerges from the previous flow of words. The dream is about to end, to use one of the most popular metaphors for narrative, and it is important that the words not take off in a new direction or unnecessarily call attention to themselves, but serve the story—the dream— as best they can.

The closing can assume several different forms, depending on the genre: it can offer a resolution or conclusion to the storyline, it can reiterate theme, it can provide a consummation to the argument, it can establish an expanded context for the piece, it can return to some integral part of the opening, it can pose some final question, it can find resonance in a quotation, it can seek closure in a vivid image or word picture. As with the opening, the closing can vary considerably in length, from a very succinct sentence or two to something quite elaborate, with most probably falling somewhere in between. Believe it or not, the closing paragraph to *Moby Dick* (not the Epilogue but the closing paragraph) consists of only one short (for Melville) thirty-eight –word sentence. Ahab and his ship have just disappeared into the Pacific. Melville writes:

> Now small fowls flew screaming over the yet yawning gulf; a sullen white surf beat against its steep sides; then all collapsed, and the great shroud of the sea rolled on as it rolled five thousand years ago.

Of course, Melville could not resist the temptation, and added an extended paragraph for an Epilogue, but the narrative has really ended here, as the sea rolls peacefully over the tormented soul of its chief protagonist Ahab, who has finally been annihilated by the nature he dared to challenge.

As we saw earlier with the restrained openings, Rick Bass is a master of brevity. Not surprisingly, he also sometimes employs very short closing paragraphs. A brief sampling:

> For a ring I gave her a pop-top from a can of Lone Star. ("Spring," an essay from *The Deer Pasture*)

In camp that night, I was the happiest I'd been yet on the trip, without really being able to define it. It was a new-start kind of happy. A Texas Hill Country deer hunting sort of happy. A third day in deer camp happy. ("Werner," an essay from *The Deer Pasture*)

It's the secret of life itself, I'll bet. I've got the next fifty years to puzzle over it. ("Autumn," an essay from *The Deer Pasture*)

The deer are secondary, they really are. ("The Storyteller," an essay from *The Deer Pasture*)

It is not the fastest route to the mountains, but it is the shortest. ("Shortest Route to the Mountains," an essay from *Wild to the Heart*)

Falling asleep is like knowing there is someone outside that understands you better than you know yourself—that quietness, that continued falling. It will not stop soon; the winter of it will not change. ("First Snow," an essay from *Wild to the Heart*.

We see a number of different approaches here, from a thematic summary in "Shortest Route to the Mountains," to an endearing romantic image in "Spring," to a philosophical observation in "Autumn," to a rhetorical repetition in "Werner." Their brevity insures that these closing paragraphs will not go unnoticed, and they take advantage of the fact that the final position within any narrative or essay is naturally emphatic.

Another contemporary nature writer who frequently employs very restrained closings is Terry Tempest Williams. Like Rick Bass, she is adept at using the sudden, compressed, or distilled closing paragraph to provide an effective resolution. These are all closing paragraphs:

I was not prepared for the loneliness that followed. ("Whimbrels," an essay from *Refuge*)

When Emily Dickinson writes, "Hope is a thing with feathers that lights upon our soul," she reminds us, as the birds do, of the liberation and pragmatism of belief. ("Pink Flamingoes," an essay from *Refuge*)

Mother and I break bread for the geese. We leave small offerings throughout the meadow. It is bread made by the monks from stone-ground grain. She puts her arm back through mine as we walk shoulder-high in sunflowers. ("Canada Geese," an essay from *Refuge*).

There are several different variations here. In the first example Williams opts for a single declarative sentence that is almost scriptural in its simplicity (another reason her language seems biblical is that she consistently employs very old, Anglo-Saxon words and also uses comparatively few modifiers). The second example includes a quotation from Emily Dickinson (an appropriate quotation used in a closing paragraph can produce a powerful resonance, as it does here). Williams chooses an image of affirmation and life—mother and daughter walking arm-in-arm through a field of sunflowers—in the last example. Not one of these closing paragraphs is longer than forty words, and yet each achieves a quiet and memorable finality.

Extremely long paragraphs of two hundred or more words can sometimes be used to achieve a particular effect in closing out a narrative. In my essay 'Radiant Darkness,' which looks at the creation of an African bushveld diorama in a natural history museum, I chose the extended paragraph approach:

Despite his premature end, the lion was, I concluded, fortunate in one respect. The relentless flow of time had ebbed and formed a quiet pool in the hall of mammals. Beyond the protective glass, the lion would never age, never fade, never know infirmity, weakness, banishment, and the final unexpected betrayals of old age. The lion would forever be in his maturity there, hunting with his pride on the plains he loved, wild and free, forever about to spring upon the unwary antelope. The sun would never go down on that day. The budding leaves would never turn brown and fall to Earth. The grass fires would never come, nor the locusts, nor the droughts, nor the bulldozed roads and mechanized plagues of man. When all his pride was dead and his progeny was scattered over the land, and even the maker of this essay was buried and forgotten, he would remain, part of a dramatic scene that might persuade futurity that paradise was worth saving. In the long darkened African hall, the restless generations slowly file past the exhibit even now, each one peering for a transfixed moment, their tired faces suddenly brightened by the sight of an old familiar home. On those green hills south of the equator the beetle-browed ancestors of the human race first stepped forward from the sheltering trees, first followed the elephant trails out into the wide open country, first stood amazed at the abundance and possibilities of life. More amazing than anything they beheld though—even those solemn appraising lions—was the instrument of their apprehension: the brain they took each night up into the branches among the stars, a swirling universe unto itself, a dominion greater than all Africa.

Here I had to slowly build a momentum through a series of both simple and complex sentences that would reach its natural culmination, rhythmically and thematically, in the thought expressed in the final phrase, "a dominion greater than all Africa." The inspiration for this paragraph was John Keats's poem "Ode to a Grecian Urn," which contrasts the permanence of art with the ephemeral nature of life. It struck me that the same could be said of the habitat groups found in a natural history museum. Like Keats's Grecian Urns in the British Art Museum, they, too, would last well beyond the lifetimes of their creators. But I wanted to take that idea one step further, and after some meditation, settled upon the "dominion greater than all Africa" concept as providing the best resolution for the essay. All of this took well over two hundred words to accomplish.

As with openings, most essay closings fall between the two extremes of very short and very long. A few examples of these more typical closing paragraphs will serve to illustrate:

> If it's wild to your own heart, protect it. Preserve it. Love it. And fight for it, and dedicate yourself to it, whether it's a mountain range, your wife, your husband, or even (heaven forbid) your job. It doesn't matter if it's wild to anyone else: if it's what makes your heart sing, if it's what makes your days soar like a hawk in the summertime, then focus on it. Because for sure, it's wild, and if it's wild, it'll mean you're still free. No matter where you are. ("River People," an essay from the book *Wild to the Heart*, by Rick Bass)

> He turns and walks his long-legged walk across the benchland. In the distance, at the pickup, an empty beer can falls on the ground when he gets in. I can hear his radio as he bumps toward town. Dust rises like an evening gown behind his truck. It flies free for a moment, then returns, leisurely, to the habitual road—that bruised string which leads to and from my heart. ("From a Sheepherder's Notebook," an essay from the book *The Solace of Open Spaces*, by Gretel Ehrlich)

> We wake at dawn to discover the desert hills shrouded in rolling clouds of vapor, seeming remote and mystical as the Mountains of the Moon. A rare and lovely sight and we are sorry to leave. We console ourselves, as we always do, with the thought that we'll be back, someday soon. We will return, someday, and when we do the gritty splendor and the complicated grandeur of Big Bend will still be here. Waiting for us. Isn't that what we always think as we hurry on, rushing toward the inane infinity of our unnamable desires? Isn't that what we always say? ("Big Bend," an essay from the book *One Life at a Time, Please*, by Edward Abbey)

52

Stars and planets, seemingly more plentiful with no competition from city lights, sparkle in the clear high-altitude atmosphere. They appear in pairs, one in the sky and one in the lake. Cassiopeia wheels up over the eastern horizon. Another world glimmers coldly down upon this one. Something splashes in the lake with a delicate neat liquid sound. Looking upward, I wonder if any of those other worlds can possibly match this one. ("The Lake," an essay from the book *Beyond the Aspen Grove*, by Ann Zwinger.)

Although these five closing paragraphs are of uniform length, the approaches could not be more different. Rick Bass chooses to repeat a key word—wild—several times to stress its close relationship to freedom, also mentioned at a key moment in the paragraph (notice also how he uses short staccato sentences to speed up the cadence). Gretel Ehrlich provides a word picture of a person that relates back to theme. Both Edward Abbey and Ann Zwinger ask a rhetorical question, although Abbey's are punctuated as such and Zwinger's is not. Both Abbey and Zwinger have also suddenly expanded their narratives with a philosophical flourish at the end.

It is sometimes advantageous to recall some aspect of your opening paragraph—a striking image, an important fact, a rhetorical question—in your closing paragraph. This can achieve a pleasing finality by giving the reader the sense that they have returned to the beginning of their travels, in the sense that any story or essay is a sort of journey of the mind. A writer who sometimes employs this technique is John McPhee, who has authored such nature books as *The Pine Barrens* and *The Survival of the Bark Canoe* and teaches a course entitled "The Literature of Fact" at Princeton University. In his extended excursion essay "The Encircled River," which appears as the first third of his book *Coming into the Country*, McPhee uses the technique with particular success. The first paragraph begins:

My bandanna is rolled on the diagonal and retains water fairly well. I keep it knotted around my head, and now and again dip it into the river. The water is forty-six degrees. Against the temples, it is refrigerant and relieving . . .

Some ninety pages later—after having floated for a number of days down the Kobuk River in northwestern Alaska—McPhee returns to this image in the closing paragraph:

> We drifted to the rip, and down it past the mutilated salmon. Then we came to another long flat surface, spraying up the light of the sun. My bandanna, around my head, was nearly dry. I took it off, and trailed it in the river.

This closing works well for two reasons. First, it recalls the central image of the opening paragraph. Second, it evokes the title of the essay "The Encircled River" (which in turn evokes the theme of natural cycles) and it further supports theme because the bandana "encircles" the author's head. I can't think of another example where a person has gotten so much good use out of a bandanna.

Another important type of closing to be aware of is the *in media res* (again, Latin for "in the middle of things") closing, which literally stops the essay in the midst of some sort of action. One of the chief practitioners of this type of closing was Ernest Hemingway, and two of the best examples are found in his novels *A Farewell to Arms* and *For Whom the Bell Tolls*, both of which include a fair amount of "nature writing" in terms of natural description. In the first instance we are left with a "postcard" sort of word picture of the protagonist leaving the hospital after his wife has died:

> But after I had got them out and shut the door and turned off the light it wasn't any good. It was like saying good-by to a statue. After a while I went out and left the hospital and walked back to the hotel in the rain.

We are watching the protagonist as he walks back to the hotel in the rain. He is no longer in the hospital, nor has he reached the hotel. He is somewhere in between. Hence, an *in media res* closing. In the second example, the protagonist is wounded and waiting for the enemy to get close enough to fire:

. . . Robert Jordan lay behind the tree, holding onto himself very carefully and delicately to keep his hands steady. He was waiting until the officer reached the sunlit place where the first trees of the pine forest joined the green slope of the meadow. He could feel his heart beating against the pine needle floor of the forest.

Here again we see the protagonist awaiting his fate, and are left to wonder at what that fate will be. The enemy is closing. He has his rifle in his arms. He can "feel his heart beating against the pine needle floor of the forest." He could live or die, but the outcome will always be uncertain (just as the outcome of life itself is always uncertain). The last image is particularly important because it recalls the first paragraph of the novel, in which Jordan is laying "flat on the brown, pine needle floor of the forest" examining an enemy position through binoculars. In both novels, Hemingway wrote and wanted to include a final chapter that explained what happened to all the characters (a typical nineteenth-century novel approach), but was persuaded not to by his editor, Maxwell Perkins at Scribner's, who believed strongly that the *in media res* closing would be more dramatic.

A variation of the *in media res* closing is the dialogue closing. Edward Abbey was fond of these, as in his essay "Running the San Juan" from *Down the River*:

> The pale sandstone bluffs come down to the water's edge; there is no shore. We're on the reservoir. I can see the high waterline on the rock: the Lake Powell Bathtub Ring. We approach a lean and lanky fellow squatting on a ledge close to the water. He's got a fishing pole beside him, a small fire going, a coffeepot, two catfish frying in a skillet. I row in close.
> "Hello there."
> "Howdy."
> "How's the fishing?"
> "Not bad."
> "Where's the crossing?"
> "You're there."

"Yeah? Where's the road?"

"About ten feet below you."

"Well, I'll be damned. Two years ago it all looked different."

"Well," he says," two years ago there was still a river here."

"Well, I'll be damned."

"Yeah," he says," it sure is."

Here the unattributed lines of dialogue with the total stranger fishing on the bank work well, particularly as they close out with the little joke on the double entendre of "dam" and "damn." We are left, as with the Hemingway closings, literally "in the middle of things," unsure where the author is going to land his boat, if he is going to land his boat. Again, that uncertainty is Abbey's point. What does the future hold for us after we dam a river?

What you want in the end is for your readers to come away with that closing paragraph forever echoing through their minds, forever reminding them of all the words that came before. While such a closing is not always possible, it should always be strived for. My closing paragraphs generally go through at least forty revisions. Seriously. I carefully weigh and balance every word and every phrase, consider the nuance of every expression, the value of even something as small as an indefinite article. I'm always looking for that striking image, that haunting phrase, that convincing fact or figure for an argument. I'm always aiming for the greatest concision and condensation possible. I'm always studying writers who I admire for their craft, for their ability to create beautiful openings and closings.

Practice Exercises

1. As in the practice exercise for the opening, select a classic work of nature writing and carefully analyze the essay closings. Are there any rhetorical paradigms that might be useful to you in an essay written, being revised, or being contemplated? Books you might wish to look at include Sigurd Olson's *The Singing Wilderness*, William O. Douglas's

My Wilderness: East to Katahdin, David Rain Wallace's *The Klamath Know*, Gretel Ehrlich's *The Solace of Open Spaces*, or Richard Nelson's *The Island Within*.

2. Similarly, take a well-known nature essay, block out the original closing, and compose a closing of your own. Create a different exit from that employed by the author. What is achieved or lost by the change? Essays that would be interesting to work with include John Muir's "The Water Ouzel" from *The Mountains of California* and Edward Hoagland's "In Praise of John Muir" from *Balancing Acts*.

3. Compose a closing for one of your essays in which you "circle back" to an important image in the opening paragraph as John McPhee did in the example provided in the chapter. Does this make for a more resonant closing?

4. Sometimes writers reach the end of their essay and find they have exhausted them, and yet they still must close the piece out. One solution to this problem is to find a quotation that fits in nicely. There are any number of quotation books available in the research section of your library, but it might be more useful to look elsewhere—in other nature books—for a quotation that achieves a particularly good effect. This was the technique I employed in the grizzly bear selection featured in *Wildlife in Peril*. For years I had carried the quotation from Faulkner's story 'The Bear' around, looking for a place to use it. I finally found it:

> A grizzly Western Slope cowboy once put it best to me, in the simple and colorful parlance of his trade: "A mountain without Old Ephraim is like a bowl of chili without the chili." Aldo Leopold put it another way, writing of these little understood animals that claim, at worst, "a cow a year and a few square miles of useless rocks." After one of the last grizzlies in Arizona, "Big Foot" of the Escudilla, was killed by a set-gun trap in a gorge, Leopold wrote that "Escudilla still hangs on the horizon, but

when you see it you no longer think of bear. It's only a mountain now." Like the loss of the fabled Mississippi black bear in William Faulkner's story "The Bear," the loss of the Escudilla grizzly corresponded with, and came to symbolize for Leopold and others the loss of the wilderness itself. The wilderness, Faulkner wrote, was something "whose edges were being constantly and punily gnawed at by men with plows and axes who feared it because it was wilderness."

The black bear of Faulkner's story was like the grizzly of Leopold's Escudilla, or the last grizzly of New Mexico killed on the Gila River in 1931, or quite possibly the last grizzly in Colorado, killed on the Navajo River on September 23, 1979:

> It loomed and towered in his dreams before he even saw the unaxed woods where it left its crooked print, shaggy, tremendous, red-eyed, not malevolent but just big, too big for the dogs which tried to ride it down for the men and the bullets they fired into it; too big for the very country which was its constricting scope . . . not . . . a mortal beast but an anachronism indomitable and invincible out of an old, dead time, a phantom, epitome and apotheosis of the old wild life . . . the old bear, solitary, indomitable, and alone; widowered, childless, and absolved of mortality— Old Priam reft of his old wife and outlived all his sons.

5. Write a closing (or, if you have time, the entire essay) in which you use a quote in some substantial way to close off the narrative. Most everyone has a favorite quotation or two the have been meditating upon for a long time—from a poem, from a scriptural or religious text, from a beloved work of fiction or non-fiction. See what you can do with this in the context of nature writing and closings.

Chapter Six

Word Pictures

"I have finished two long short stories, one of them not
much good and finished the long one I worked on before I
went to Spain ["Big Two-Hearted River"] where I'm
trying to do the country like Cezanne."

— Ernest Hemingway,
letter to Gertrude Stein and Alice B. Toklas,
August 15, 1924

Comprehensive word pictures, running over many pages, were once commonplace in nature literature. Read the prose of Washington Irving, describing the Oklahoma grasslands in *A Tour of the Prairies* (1835), or the prose of Henry Stanley, describing the tropical rainforest of the Congo Basin in central Africa in his book *In Darkest Africa* (1890), and you read descriptive nature writing from this period. Nature writers in the nineteenth-century sought to achieve with words and paper what the respective artists of their times, the illuminists of the Hudson River School and post-Romantics of the American far west (Hayden and Moran), were accomplishing with oils and canvas. Here, for example, is part of the almost grotesquely detailed passage just referred to in Stanley:

Imagine the whole of France and the Iberian peninsula closely packed with trees varying from 20 to 180 feet high, whose crowns of foliage interlace and prevent any view of sky and sun, and each tree from a few inches to four feet in diameter. Then from tree to tree run cables from two inches to fifteen inches in diameter, up and down in loops and festoons and W's and badly-formed M's; fold them round the trees in great tight coils, until they have run up the entire height, like endless anacondas; let them flower and leaf luxuriantly, and mix above with the foliage of the trees to hide the sun, then from the highest branches let fall the ends of the cables reaching near to the ground by hundreds with frayed extremities, for these represent the air roots of the Epiphytes; let slender cords hang down also in tassles with open thread-work at the ends. Work others through and through these as confusedly as possible, and pendent from branch to branch—with absolute disregard of material, and at every fork and on every horizontal branch cabbage-like lichens of the largest kind, and broad spear-leaved plants— these would represent the elephant-eared plant—and orchids and clusters of vegetable marvels, and a drapery of delicate ferns which abound. Now cover tree, branch, twig, and creeper with a thick moss like a green fur. Where the forest is compact as described above, we may not do more than cover the ground closely with a thick crop of phrynia, and amoma, and dwarf bush; but if the lightning, as infrequently happens, has severed the crown of a proud tree, and let in the sunlight, or split a giant down to its roots, or scorched it dead, or a tornado has been uprooting a few trees, then the race for air and light has caused a multitude of baby trees to rush upward— crowded, crushing, and treading upon and strangling one another, until the whole is impervious bush.

This goes on for twelve more pages.

Just as art has radically changed in the twentieth century, so has the practice of drawing with words. Hemingway, for example, practiced minimalism, and, following his Post-Impressionistic mentor Paul Cézanne, drew his word pictures in as little space as possible. His influence is seen most directly in then-contemporary nature writers Bob Marshall, Aldo Leopold, and Joseph Wood Krutch. Aldo Leopold, for example, described the top of White Mountain in southern Arizona as follows:

> The top of the mountain was a great meadow, half a day's ride across, but do not picture it as a single amphitheater of grass, hedged in by a wall of pines. The edges of that meadow were scrolled, curled, and crenulated with an infinity of bays and coves, points and stringers, peninsulas and parks, each one of which differed from all the rest. No man knew them all, and every day's ride offered a gambler's chance of finding a new one. I say 'new' because one often had the feeling, riding into some flower-spangled cove, that if anyone had ever been here before, he must of necessity have sung a song, or written a poem.

Washington Irving or Henry Stanley (not to mention John Muir or John Burroughs), by contrast, would probably have taken a page or two to describe the same scene. It is doubtful that the elaborate descriptions from the last century will come back into favor with readers in the near future. Until such time as they do, writers (at least those interested in being published and read) should take into consideration several important points.

There are, to begin with, at least two important historical reasons for the change from the lengthy descriptions of the nineteenth century to the more restrained descriptions of the twentieth century. First, as field cameras became lighter and their use more widespread in the second half of the nineteenth century, whether by Matthew Brady in the Civil War or by William Henry Jackson in Colorado in the 1870s, descriptive prose became increasingly challenged by the photographic image. Eventually, nature writers began to include photographs in their books, as Theodore Roosevelt did in the 1890s, which reduced the writer's responsibility to describe through word pictures. Today, of course, descriptive prose must also compete with the moving visual image: film, television, and videotape. Second, we live in a faster paced world in the 1990s, and the modern reader is less tolerant of lengthy descriptions than the nineteenth century reader, for whom the distractions of television and telephone, automobile and jet plane, radio and CD player did not exist. Mark Twain could take an entire chapter to describe the Mississippi River in *Life on the Mississippi* (1883). Modern writers would probably find themselves with a lot of unread books and critical reviews if they attempted the same.

This does not mean, however, that word pictures are a thing of the past and should be foregone in the interest of a fast-paced narrative full of action and compelling character revelation. To the contrary, word pictures are one of the basic structural units of narrative and essential to your task as a writer, which is to help the reader visualize the world you are creating with words. Having said this, word pictures generally fall into three broad categories: 1) they describe a natural scene (either static, like a snow-bound winter valley, or dynamic, like a volcanic eruption); 2) they describe a person who is central to the narrative (marginal characters need not be comprehensively described); or 3) they describe some sort of pivotal or transitional action (such as a person surfing, a person tranquilizing an elephant, a person rafting).

Most often, you will find yourself working within the first category, i.e.; painting a word picture of a landscape. One possible approach is to begin with the foreground, move to the mid-ground, and then conclude with the background. This conventional technique has been taught in basic drawing classes since the age of the pharaohs. When you have mastered this paradigm, you can feel free to experiment with other, more novel approaches to the descriptive process. In fact, experiment as much you can, because only in that way will you develop an individual style (something we will talk about in a future chapter). Above all, look for texture, color, lines, shadowing, and movement in the same discerning way that an artist does. Search for those important details that make the scene singular. At the same time look for images that will make the scene familiar to your readers, both in the present time and in the future time (always remember to write for the future audience, for the person reading your book not tomorrow but a century from tomorrow). Sometimes it helps to actually sketch the scene on a piece of paper before beginning the process of committing it to paper with words.

Other times it can be beneficial to close your eyes and try to describe the scene to yourself from memory, again before beginning the writing process.

Jim Harrison is particularly adept at landscape word pictures, as in the opening paragraph to his fishing essay "A Day in May:"

> Without having flown over this particular stretch of water southwest of Key West, I can still envision it topographically: the infinite shadings of blue over the tidal flats—azure, indigo and the predominant light turquoise of the shallows with the paler striations of white sand. Then the brown turtle grass, the dark outlines of coral outcroppings or tidal cuts that game fish use to reach the feeding grounds, and the darker green random splotches of mangrove keys. Farther to the south is a sometimes garish penumbra of purple, that imaginary point where the Gulf and the Atlantic meet in a great ocean river, the Gulf Stream.

Here, within the span of one paragraph and eighty words, the author has effectively painted a large, nearly epic scene of tropical grandeur. He has begun in the foreground, with the "shadings of blue over the tidal flats," moved to "the dark outlines of coral outcroppings," and finished with the "sometimes garish penumbra of purple" that is the Gulf Stream. Harrison has relied primarily on colors—blue, azure, indigo, turquoise, white, brown, green, purple—to portray this bright world of Winslow Homer and Jimmy Buffet for his reader.

The second broad category involves word pictures of people. In some ways, these are the most difficult word pictures; our familiarity with the human form has led us to forget to closely observe the nuances that make a particular person unique and interesting. Certain writers have a gift for these. John McPhee is one of those. In his book *Coming into the Country*, for example, he deftly brings people to life in a sentence or two. Here he introduces us to trapper Dick Cook of Eagle, Alaska:

> Cook is somewhat below the threshold of slender. He is fatless. His figure is a little stooped, unprepossessing, but his

legs and arms are strong beyond the mere requirements of the athlete. He looks like a scarecrow made of cables. All his features are feral—his chin, his nose, his dark eyes. His hair, which is nearly black, has gone far from his forehead. His scalp is bare all the way back to, more or less, his north pole. The growth beyond—dense, streaked with gray— cantilevers to the sides in unbarbered profusion, so that his own hair appears to be a parka ruff. His voice is soft, gentle—his words polite. When he is being pedagogical, the voice goes up several registers, and becomes hortative and sharp. He is not infrequently pedagogical.

Notice how McPhee uses words and imagery like "feral" and "north pole" and "parka ruff" to tie his subject to his surroundings. McPhee has also employed what is known as an emphatic interruption (a phrase set inside double hyphens) in the phrase "dense, streaked with gray" to stress the fact that Cook's hair may be absent on top but is thick beyond. It is prudent to use these emphatic interruptions, as with any rhetorical device, judiciously. McPhee's reference to Cook's voice, and speaking habits, helps to bring the character further to life. Interestingly enough, several years ago my father and I met Dick Cook at the Murie Cabin along the East Fork of the Toklat River in Denali National Park. Cook had been hired by the park service to refurbish the fifty year old cabin, which is still used by wolf and bear researchers. I can vouch for the fact that, nearly twenty years after it was written, McPhee's word picture of Cook is still accurate. Cook looks the same, except that the black hair is now a sort of blondish-white.

Finally, we have word pictures of action. What is most important here is what is left out. Don't overwhelm your reader. Just give them the essence. If you stop to describe every frame, the action will stop. Your chief concern must be with motion. You may wish to use shorter sentences, as I have in this paragraph, because staccato sentences cause the cadence to speed up, whereas periodic sentences cause the tempo to slow down. Here, for example, is a scene from Edward Abbey's *Desert Solitaire* in which a man is about to die. The man is in a truck rolling toward the side of a canyon:

Thinking carefully, Mr. Graham decided to ease the truck down close to the edge of the mesa, stop it there and toss a match in on his partner before pushing the whole works over. The truck as parked parallel to the rim, not facing it, so Mr. Graham after removing the rocks from in front of the wheels climbed into the driver's seat, pushing Husk's legs out of the way, and automatically, out of habit, turned on the switch. He had almost stepped on the starter before he realized the danger. Without starting the motor he disengaged the clutch, took the truck out of gear, and turned the wheels downhill. There was an awful lot of loose play in the steering. As the truck began to roll Mr. Graham's right foot groped for the brake pedal, found it and pushed it down to the floorboard without meeting the slightest hint of resistance. In sudden alarm he grabbed for the parking brake and found the handle missing. There was nothing there. All at once Mr. Graham knew that more than anything else in the world he wanted to get out of that truck (from the essay "Rocks" in *Desert Solitaire*).

The movement here is not only fast, it is relentless. As the tension builds, so do the sentences shorten, and quicken ("There was nothing there."), until, in the next paragraph, Mr. Graham is seen flying through space with a fleeting view of "the sky, a few stars, the tranquil moon" before gravity puts him on the ground.

In his "Treatise on Painting" Leonardo da Vinci wrote that "Painting is poetry that is seen rather than felt, and poetry is painting that is felt rather than seen." The latter is especially true for nature writing, which by virtue of its subject has a lyric, poetic quality. Your aim is to paint word pictures that cause the reader to "feel" a scene that cannot, in the strictest sense, ever be viewed. To do this properly, you cannot paint every detail, rather you must find what is essential. In the hands of a master, a word picture is something to behold. Listen to Clarence Dutton describe a sunset over the Grand Canyon:

At length the sun sinks and the colors cease to burn. The abyss lapses back into repose. But its glory mounts upward and diffuses itself in the sky above. Long streamers of rosy light, rayed out from the west, cross the firmament and converge again in the east, ending in a pale rosy arch, which rises like a low aurora just above the eastern horizon. Below it is the dead gray shadow of the world. Higher and higher climbs the arch, followed by the darkening pall of gray, and as it ascends it fades and disappears, leaving no color except the afterglow of the western clouds and the lusterless red of the chasm below. Within the abyss the darkness gathers. Gradually the shades deepen and ascend, hiding the opposite wall and enveloping the great temples. For a few moments the summits of these majestic piles seem to float upon a sea of blackness, then vanish in the darkness, and, wrapped in the impenetrable mantle of the night, they await the glory of the coming dawn.

All the wonder and exultation of day's end in the Painted Desert are summoned up here. Nothing has been left out that was necessary, nor has anything necessary been unnecessarily adorned or added to. Such is the power of the artist in the writer, when, through craft and experience, inspiration and discipline, language is made to draw pictures.

Practice Exercises

1. In his book *Arctic Dreams* Barry Lopez describes an oil painting entitled "The Icebergs" by Frederic Edwin Church (1826–1900) in the following paragraph:

   The oil painting he [Church] produced from these sketches came to be called "The Icebergs." It is so imposing—6 feet by 10 feet wide—a viewer feels he can almost step into it, which was Church's intent. In the foreground is a shelf of ice, part of an iceberg that fills most of the painting and which rises abruptly in the left foreground. On the right, the flooded ice shelf becomes part of a wave-carved grotto. In the central middle ground is a becalmed embayment, opening onto darker ocean waters to the left, which continue to a stormy horizon and other, distant icebergs. Dominating the background on the far side of the embayment is a high wall of ice and snow that carries all the way to the right of the painting. In the ocean air above is a rolling mist. The shading and forms of the icebergs are expertly limned—Church was an avid naturalist, and conscientious about such accuracy—and the colors, though slightly embellished, are true (220).

Choose a favorite photograph, like Ansel Adams's *Moonrise over Hernandez, New Mexico*, or a favorite painting, like Charles Russell's *A Wounded Grizzly*, and paint a word picture of the visual image, utilizing the same paradigm (foreground, mid-ground, background) as Lopez in his passage. Try to aim for at least 150 words and to focus on those details that give the natural image its particular life.

2. Visit your local zoo or aquarium and choose your favorite animal. Describe not only the animal, whether it be a gorilla or a killer whale, but also the manner in which the people respond to the animal.

3. Most state game and fish or wildlife departments, not to mention museums, have a large collection of animal skulls and skins that can be viewed and, in some cases, borrowed by the general public, and by educators. Obtain either a striking skin, such as the hide of a wolf, or a beautiful skull, such as the skull of a mountain lion, and paint a word picture of it. This exercise is particularly useful if figurative language (metaphors and similes) are also utilized. For example, in my book *Wild Africa* (1992) I described the skin of an African lion as follows:

After completing the measurements I inventoried the natural damage to the hide. This lion had evidently led an active life, for the skin was a veritable catalogue of old injuries. The lion's left ear was badly tattered, his right ear was split part way down the middle, the black tail tuft was all but gone, and a myriad minor scars and wounds decorated the grizzled white muzzle and black nose. All of these were old, well- healed injuries. The back of the lion, behind the mane, bore evidence of some terrific wounds—long deep gashes and punctures—probably inflicted by the death struggles of his horned and hoofed victims. In other places you could see where the claws of other lions, likely contending males, had cut deep into the skin. The parallel claw marks gave the appearance of the dried migratory trails of wildebeests over muddy grasslands. There was a single bullet wound forward on the left shoulder, and there was no exit wound, indicating the 250-grain magnum round had either fragmented after entry or impacted a solid structure and remained there. This death, though sudden and violent, would have been far more merciful than those inflicted by nature according to most studies. The colors of the skin were the natural colors of the earth—a yellow sandstone covered most of the upper portions of the body, fading to a pale white limestone on the belly and the undersides of the legs. The dark brown mane was quite impressive, and thickly covered the shoulders, neck, and head like a tangled thornbrush woodland in the dry season. The markings on the face were more complex, designed as they were to accentuate the whole range of emotions. Dead ticks still clung tenaciously to the long mane hairs where they had madly scrambled after the insecticide was applied. These loathsome parasites, a natural vector for disease, bedevil almost all warm-blooded veretebrates of Subsaharan Africa.

4. Choose a favorite outdoor scene, a streamside or a forest clearing or a mountaintop, and sit there for as long as it takes for you to absorb its general shape and salient details. When you have studied the scene long enough to close your eyes and describe it accurately to yourself, pick up a pencil and paper and roughly draw it. You don't have to be a great artist to do this. The point of the exercise is to get you inside the scene in a way that passive observation does not, to give you a sense of the textures, of the shadowing, of the lines that may not be as evident in a few glances around the country. When you have completed all of this, paint a word picture of it. Try to use the foreground, mid-ground, background paradigm.

5. Choose a friend or an acquaintance and paint a word picture of them. Focus on the qualities that make them unique, but also try to universalize whatever is particular in them. Remember the importance of eyes— not only their color and physical characteristics, but also whether they appear to be sad or happy, confident or uncertain, tired or full of energy. This is a good exercise for metaphors and similes. Few things make nature essays come to life more quickly than the appearance of people, especially people who are well described and become as familiar to the reader as they were to the author. Here, for example, is Francis Parkman describing a Kansas Indian chief he met along the Oregon Trail in 1846. Though a century-and-a-half old, the sixty-one–word text describes the man so well we can immediately form an image of him:

> His head was shaved and painted red, and from the tuft of hair remaining on the crown dangled several eagle's feathers, and the tails of two or three rattlesnakes. His cheeks, too, were daubed with vermillion; his ears were adorned with green glass pendants; a collar of grizzly bears' claws surrounded his neck, and several large neckles hung on his breast.

6. Observe a complicated scene in nature, such as a flight of geese approaching and landing on a woodland lake, or a herd of elk feeding

at timberline, or a kayaker coursing through some tough rapids, and describe that scene, using a minimal number of words (especially a minimal number of modifiers and intensifiers). Aim for a prose that is as lean and muscular and nature herself, and that expresses the action with as much economy, and energy, as possible.

7. Sometimes in the course of your writing you will be called upon to paint a word picture of a season or a particular month of the year. This is always a challenge, as seasonal changes are often quite sweeping and you must aim for those particular images and details that evoke or summon up a vast canvas. Pick your favorite month or season, and write an extended word picture of 100–150 words. To help you get started, here is a word picture of the month of September in northern Alaska from my book *Out Among the Wolves* (1993).

> September is a time of rapid change on the Alaskan tundra, as the first strong arctic fronts sweep in from the Bering Sea and Siberia beyond, carrying with them sub-freezing temperatures, permanent snow, and the last of the migrating geese, ducks, and cranes. In many ways, this is the most beautiful season of the year, as the lowering termination line of the previous night's snowfall contrasts sharply with the rainbow colors ripening on the autumn tundra—the warm salmon reds of the blueberry patches, the soft yellows of the willow thickets, the orange stands of dwarf birch, the dark honey of the sedge hummocks, the bright green of the scattered spruce islands, and the worn saddle hue of the long, windswept grass hills. The stars return to the sky after a four-month sabbatical and some nights the northern lights are wonderful. In the high mountains the caribou pour through the passes by the hundreds, heading toward distant pastures on their strangely kicking hooves. The grizzlies are stunning in their shaggy winter coats, and so are the wolves, dressed in those thick capes so prized by native

trappers, who sew the ruffs to their parkas in order to keep the frost off at fifty below. The bull moose rub the velvet from their antlers and begin to drift out of the peaks into the valleys searching for cows. Down along the rivers, rafts of ice fall through the rapids and gather in pools. Withered stalks of fireweed empty their feathered seeds to the wind, and merlins (*Falco columbarius*) fly low in amted pairs over the mosquito-less bogs, looking for one last mouse before turning south to Oregon. Sometimes you see a wolverine as it shuffles out on the sandbars and searches the bleached piles of driftwood for old salmon. The whole country is restless. For the wolves, this is a particularly difficult time, as resources begin to dwindle and competition within and between the packs increases. This is the time of highest mortality for the year's young.

Chapter Seven

Figurative Language

By far, the greatest power of all is to be a master
of the metaphor. It is the one thing that cannot
be learned from others; and it is also a sign of genius,
since a good metaphor implies an intuitive perception
of the similarity in dissimilars.

— Aristotle, *Poetics*

The pomegranate is cultivated here.
The fruit is about the size of an orange, has at thick,
tough skin, and when opened resembles a many-
chambered box full of translucent purple candies.

— John Muir,
A Thousand-Mile Walk to the Gulf

The two figures of speech most often employed by nature writers are metaphors and similes. The word metaphor comes from the Greek μετα meaning "to bear, to carry." A metaphor does just that. It causes a word "to carry" a second meaning, so that the word has an additional idea superimposed on what it normally represents. For example, it is one thing to say that "the ridgeline was as steep as the back of a wolf." It is another to say that "the ridgeline was wolf-backed." In the first instance, the ridgeline is being compared, through the adverb "as," to the back of a wolf. In the second, the ridgeline has quite literally become "wolf-backed." Similes (from the Latin *similis*, or like) use either "like" or "as" to draw the basis of comparison, as we saw in the first example. Aristotle believed that effective similes "give an effect of brilliance" but preferred the metaphor to the simile because metaphors are less grammatically complicated and because metaphors actually say that "this is that" and create a stronger impression with readers.

Samuel Johnson once suggested that "a simile may be compared to lines converging at a point, and is more excellent as the lines approach from a greater distance." This is true only to a point—if the line is too long it breaks, and we say that the figure of speech has become over-reaching. Metaphors and similes can in unpracticed hands cause substantial damage to structures of meaning, somewhat like power tools misused by the unskilled. Writers should avoid clichés ("he let the cat out of the bag"), mixed metaphors ("he was a jungle cat pacing on the pressure cooker of the sidelines"), and over-reaching or strained metaphors ("the sky was as red as freshly-spilled blood"). In the hands of an accomplished writer, however, the effect of such literary devices as metaphors and similes can be dazzling, as in this passage from Barry Lopez's fine essay "Drought":

> The river has come back to fit between its banks. To stick your hands into the river is to feel the cords that bind the earth together in one piece. The sound of it at a distance is like wild horses in a canyon, going sure-footed away from the smell of a cougar come to them faintly on the wind.

In this case, we see the simile conveys a double meaning with poetic effect—that the rainwater restoring the river after a long drought has brought new life to the canyon even as it has reinvigorated the imagination of the artist who lives there. The fact that this is the final paragraph in the essay on drought gives added resonance to the comparison, and makes the ending all that more powerful.

I first learned about metaphors and similes at an early age, sixteen, while taking a summer writing workshop (see Chapter 15 on Work-shopping). My teacher was a poet, and he impressed upon us the importance of using these tools as soon as possible in our writing. Toward this goal, he had us maintain a longbook in which we copied down the most effective images, metaphors, and similes encountered in our readings of poetry and prose. We then emulated these rhetorical paradigms. What I found then as a student and now as a teacher is that Aristotle is absolutely wrong when he says, referring to the metaphoric mode of thinking, that "it is the one thing that cannot be learned from others." When novice writers are provided with examples, such as are found in the practice section for this chapter, and given time to experiment, take risks, fail, learn and grow, they can produce metaphors and similes of great originality and force. Every spring semester in my nonfiction workshop I present students with exercises similar to those found with this chapter, and by the end of the week it is positively remarkable to see the growth curve. One of the axioms of education is that people become what you tell them they are. If you follow Aristotle's dictum—that metaphors cannot be taught—and you sadly inform students that they are either born or not born with this gift, the class will dejectedly produce insipid clichés, clashing metaphors and strained figures of speech. If, on the other hand, you wisely ignore Aristotle's assumption, and empower people with the exciting truth about the magnitude of their natural abilities, you will be treated with a response to the exercise that is nothing short of astonishing.

One of the earliest and most extensive uses of metaphors and similes in Western Literature occurred in *The Iliad*, which was composed by Homer, a Greek poet of coastal Asia Minor, in the eighth century BC. This was the primary literary work from which Henry David Thoreau, the first American nature writer to make a contribution to world literature, learned about figurative language, particularly extended similes. Almost all of the metaphoric imagery employed in *The Iliad* derives from pastoral nature and is skillfully placed in the narrative at key points to counterpoint the chaos and violence of war with order, peace, and harmony. Two representative examples from the over 160 extended similes found in the poem will illustrate:

As when some Maionian woman or Karian with purple colors ivory, to make it a cheek piece for horses; it lies away in an inner room, and many a rider longs to have it, but it is laid up to be a king's treasure, so, Menelaus, your shapely thighs were stained with the color of blood, and your legs also and the ankles beneath them (Book IV).

He dropped them to the ground in the dust, like some black poplar, which in the land low-lying about a great marsh grows smoothtrimmed yet with branches growing at the uttermost tree-top: one whom a man, a maker of chariots, fells with the shining iron, to bend it into a wheel for a fine-wrought chariot, and the tree lies hardening by the banks of the river; Such was Anthemion's son Simoeisios, whom illustrious Ajax killed (Book IV).

In the first case we have a simile that makes a small point of comparison between dissimilar objects—the dye on the ivory horse cheekpiece is like the blood on the leg of the Greek general Menelaus (estranged husband of Helen). In the second case we have a simile that explores the matrix of correspondences created by the analogy—the death of one man (Simoeisios), strikingly compared to the felling of the poplar for the chariot wheel, is used to evoke the death of many at the hands of the Greek warrior Ajax. Both extended similes counter the violent world of a war on alien shores with the peaceful cosmos of nature and home. Novice nature writers would be well advised to follow in Thoreau's footsteps, and undertake a careful study of these two poems in order to improve their own writing (I might also strongly suggest Virgil's *Aeneid*, which similarly uses bucolic nature imagery, via extended similes, to counterbalance and quietly comment upon the same anti-war theme).

Thoreau's most celebrated use of figurative language occurs in "Spring," the next to last chapter in *Walden*. In this passage—known popularly as the "clay bank" passage—the author describes a deep cut in a hillside made by railroad engineers. Thoreau is fascinated by the frozen soil as it begins to thaw, melt, and finally flow down the slope in the warmth of the sun. Like the Roman Stoics whom he closely studied, Thoreau had concluded that all the earth—animate and inanimate alike—is alive, and wrote early in the clay bank passage that "There is nothing inorganic. The earth is not, then, a mere fragment of dead history, strata upon strata, like the leaves of a book, an object for a museum and an antiquarian, but living poetry, like the leaves of a tree—not a fossil earth, but a living specimen." After this initial observation, more similes and metaphors crowd upon the page with an Ovidian exhuberance:

> When the frost comes out in the spring . . . the sand begins to flow down the slopes, like lava . . . Innumerable little streams overlap and interlace with another . . . As it flows [the liquified earth] takes the forms of sappy leaves or vines . . . and ressembling . . . the lacinated, lobed and imbricated thalluses of some lichens; or you are reminded of coral, of leopard's paws orbirds' feet, of brains or lungs or bowels, and excrements of all kinds. It is a truly grotesque vegetation, whose forms and color we see imitated in bronze, a sort of architectural foliage more ancient and typical than acanthus, chickory, ivy, vine, or any vegetable leaves . . . The whole cut impressed me as if it were a cave with its stalactites laid open to the light. . . . When the sun withdraws the sand ceases to flow, but in the morning the streams will start once more and branch and branch again into a myriad of others. You here see perchance how blood-vessels are formed. If you look closely you observe that first there pushes forward from the thawing mass a stream of softened sand with a drop-like point, like the ball of the finger, feeling its way slowly and blindly downward . . . What is man but a mass of thawing clay?

In this intensely metaphoric passage, the death of his brother John—remembering that John cut the "ball of his finger" off while shaving with a straight razor and subsequently died from tetanus—is finally put to rest in the revelation that all things have life, even the half-frozen earth oozing forth from the grave of winter. Such are the power of metaphors and similes to reconcile disparities and transform human consciousness.

Figurative language continues to pervade nature writing today. In Barry Lopez's *Arctic Dreams*, for example—a book selected almost at random from the crowded shelves above my computer table—I open the first page and find five similes in the first three hundred words of the Preface. Lopez is camped in Arctic Alaska and presents the reader with a vivid word picture of the environs. It is still summer and the migrating caribou "drift like smoke through the valley." The "late-night sun [is] small as a kite in the northern sky." The horned lark sitting on its nests stares "back resolute as iron." The eggs of the golden plovers glow "with a soft, pure light, like the window light in a Vermeer painting." The Lapland longspurs are "still on their nests as stones, their dark eyes gleaming." Each of these five similes effectively helps the reader to visualize the scene, with comparisons to drifting smoke, a high-flying kite, the element iron, the light in a Vermeer painting, and small stones. Elsewhere in the book Lopez works, as did Homer and Thoreau in the examples provided above, with more extensive figures of speech, as in this lengthy description of a musk ox in which he strives for a cumulative effect of dignity and balance:

> A lone animal emerged from tall, dry grasses at the foot of a slope below me. The grasses rolled in his wake, until he stood stolid on open ground, his long flank hair falling still with the stilling grass. In that moment I was struck by qualities of the animal that have stayed with me the longest: the movement was Oriental, and the pose one of meditation. The animal seemed to quiver with attention before he lowered his massive head and moved on, with the most deliberate step I have ever seen a large animal take. The shaft of a dark horn came into view, forward of the high shoulders and the full collar of his distinctive mane. The muskox settled then in my mind as a Buddhist monk, a samurai warrior. In the months after, these characteristics proved impetuous; but like many unbidden insights they served and I retain them.

The musk oxen stands "stolid on open ground" in a pose "of meditation" with a presence like "a Buddhist monk" or a "samurai warrior"—all of this interacts to produce the desired image of restrained power and deliberate movement.

Another contemporary author who is quite skilled in the use of metaphors and similes is Linda Hasselstrom, who lives on a working cattle ranch near Rapid City, South Dakota. Listen to this passage, for example, from her essay "Finding Buffalo Berries" in *Land Circle*:

> The jelly is a tawny peach color, and the flavor is hard to describe. I might compare it to apple pie with lemon: sweet, extra tangy. But another element lurks in the flavor that I can't compare to anything else. I think it's the essence of wildness, clean prairie air made solid. It contains the deer that nibbled the leaves in winter, the brush of a grouse's wing as it picked berries from the ground, the blundering invulnerability of a porcupine living under the ledge. It's the taste of blinding white drifts slowly being built and smoothed into glittering sculpture outside the house as you make morning toast, slathering it with butter and buffalo berry jelly. The jelly brings the flavor of summer heat to your tongue, a sheen of sweat to your shoulders; even as you watch the blizzard, it reminds you of spring fragrance and the cool nights of fall.

What a delightful passage we have here, and one that vividly comes to life with the many metaphors—the flavor of the buffalo berry is like "the essence of winter" or "the clean prairie air made solid." It harbors the lingering aftereffects of the nibbling deer, the pecking grouse, and the blundering porcupine, not to mention "the taste of blinding white [snow] drifts" and "the flavor of summer heat." In just one hundred words Linda Hasselstrom, ruminating over berries, summons up the whole beauty and grandeur of the Great Plains, and it is doubtful she could have accomplished this feat without employing metaphors.

I frequently employ both metaphors and similes in my nature essays, probably because I spent ten years writing poetry before turning to prose, and such tools are more essential in lyric poetry, with the emphasis on concision and ironic effect, than in prose narration. In the essay "Summer Solstice," for example, which chronicles a summer's longest day spent in the shadow of Denali, I describe the mountain in the first paragraph as "a massive peak as old as the rivers and deeply scarred, with a scattering of clouds downwind like a school of arctic grayling holding in the pool behind a boulder." In earlier drafts of this paragraph, the clouds were compared to boats holding in anchorage near an island. I made the change because the grayling are an indigenous species and also evoked a clearer visual image of the scene, with the high altitude winds sweeping around the mountain and leaving the air in the lee as a quiet pool for the elongated clouds to form in. About midway through the essay I returned to the island-image in the following extended simile:

> Because of the increasing obscurity of the lowlands [in the summer twilight] Denali at this hour begins to resemble a colossal island, with the green foothills, towering over dim lower terrain in the same way the steep headlands of an island plunge dramatically into the gray sea. The clouds float off the headlands like enormous ships riding at quiet anchorage. From this hill I gaze up at the mountain as a man in a tiny boat might look up at a vast island from the momentary height of a passing wave, wondering where it is safe to put in and what sort of people live there and what language they speak.

I always save discarded metaphors or similes while writing an essay, because, like a driveway mechanic working on a car engine, I never know when I might later need a salvageable part.

Toward the end of the solstice essay, the sun is down and the light is beginning to go off the mountain. The lenticular clouds are still there and "the moon . . . is sinking fast through [them], like the freshly laid egg of a trumpeter swan in the thick white feathers of a nest in the reeds." Again, I chose an indigenous species for the comparison, and focused on the egg for the simile because the shape is similar to that of the moon and because the egg evoked the whole sense of rebirth associated with the solstice fast (remembering again that the word nature comes from the Latin root *nasci*, to be born). The essay concludes with an extended simile that compares the twilight mountain to a sleeping child, and again emphasizes the rebirth motif that pervades the narrative:

> A boreal owl calls to me from the edge of the high spruce forest, and I turn to gather my day pack and walking stick. The grade is steep and I grasp the alpine willow branches tightly with my free hand to avoid slipping down the slope, as a child still learning to walk holds his mother's hand. I stop halfway for one last look. Just when I was certain the spectacle could not be surpassed, the mountain has become even more resplendent. The snowy bulk is luminous with the soft colors of the arctic evening, all those lavenders and lilacs, violets and blues, and resembles a giant piece of fluorite glowing beneath a fluorescent lamp. I am not worthy of such beauty. Consider this. Each night I read to my infant son before he goes to sleep. For a long time, as I read him stories of bears and bumblebees, birch trees and butterflies, he is wide awake and there is that lively twinkle in his eyes and that quick smile each time he sees me smile. And then gradually his eyelids become heavy and his head nods and he falls asleep. I turn at the nursery door and, just before the lights go off, see that he is even more beautiful now, at rest, than when awake, with his face relaxed, his eyes closed, his little hands at rest over his heart, asleep in gentle dreams. The mountain is most beautiful when it dreams.

When the Chinese philosopher Confucius wrote in an often-quoted passage that the teacher takes a raw piece of jade—the student's mind—and turns it into a work of art—the educated person, he was expressing perfectly the primal relationship of metaphor to human life. Metaphors are the means by which we express ourselves (all language is metaphoric in that words signify things), make arguments, analyze contrasting ideas, and explore the fundamental unity of the universe. Those nature writers who acquire facility with the metaphor and the simile will have a distinct advantage in expressing themselves over those that do not. The British critic John Middleton Murry expressed that fundamental importance eloquently in his seminal book *Countries of the Mind* earlier in the century:

> The investigation of metaphor is curiously like the investigation of any of the primary data of consciousness: . . . Metaphor is as ultimate as speech itself, and speech as ultimate as thought. If we try to penetrate them beyond a certain point, we find ourselves questioning the very faculty and instrument with which we are trying to penetrate them.

The Romantic poet Percy Byshe Shelley offered an equally cogent observation:

> Language is vitally metaphorical; that is, it marks the before unapprehended relations of things . . . If no new [writers] should arise to create afresh the [metaphoric] associations . . . language will be dead to all the nobler purposes of human intercourse.

Indeed, part of our responsibility as writers, especially as nature writers, is to break new ground, to create new metaphors and similes as well as new themes and styles, and to provide readers, as Murry and Shelley intimate, with fresh ways of looking at the world and thinking about the world.

Practice Exercises

1. Read through a nature book in which figurative language is used extensively, such as Annie Dillard's *The Pilgrim of Tinker Creek* or Gretel Ehrlich's *The Solace of Open Spaces*. Keep a longbook in which you enter all of your favorite metaphors and similes. Study these, as rhetorical paradigms, and try to replicate their successes in your own work.

2. Working alone or with a group of fellow nature writers, use the following phrases to create several fresh similes. Beware of clichés. Provide several possibilities for each phrase. Then use five of your best similes in sentences. In one example you might want to try for a humorous or satiric effect.

Example: as red as . . . (response—"as red as the little crimson spots on the back of a brook trout," "as red as an Arizona sunrise," "as red as the feathers on a cardinal")

   1. as dark as . . .

   2. as smooth as . . .

   3. as wrinkled as . . .

   4. as bright as . . .

   5. as green as . . .

3. In this exercise we will attempt to change literal language into figurative language. In each sentence, the underlined word is used literally. Write a sentence in which the same word is used figuratively.

Example: The aspen leaves trembled in the wind. (response—"The northern lights trembled in the Alaskan night sky like the thoughts of a mother sleepless with concern for her baby.")

   1. The porpoises swam beside the boat.

   2. The flowers drooped under the hailstorm.

   3. The volcano exploded quite suddenly.

   4. The young grizzly bear cubs played in the grass.

5. The eagle soared over the plain.

4. In this exercise we will develop a paragraph from an analogy used in one of the following phrases.

   Example:_____ is like a welcome rain after a long period of drought. ("Love is like a welcome rain after a long period of drought . . . ")

   1. _____ is like a river slowly carving a canyon.

   2. _____ is like a trail winding through the mountains.

   3. Fishing for trout in a new river is like
      _____.

   4. Damming a wild river is like_____.

   5. Peering through a microscope at one-celled organisms is like_____.

5. The following is an excerpt from Henry David Thoreau's well-known essay "Ktaadn," which relates the ascent of the tallest mountain in Maine. Pay particular attention here to the similes Thoreau employs to paint a picture of the scene. Climb a prominent eminence near your home, whether you live in Honolulu, Phoenix, or Richmond, and compose a description which uses figurative language to more vividly portray the scene from the top.

   From this elevation, just on the skirts of the clouds, we could overlook the country, west and south, for a hundred miles. There it was, the State of Maine, which we had seen on the map, but not much like that—immeasurable forest for the sun to shine on, that eastern stuff we hear of in Massachusetts. No clearing, no house. It did not look as if a solitary traveler had cut so much as a walking-stick there. Countless lakes—Moosehead in the southwest, forty miles long by ten wide, like a gleaming silver platter at the end of the table; Chesuncook, eighteen long by three wide, without an island; Millinocket, on the south, with its hundred islands; and a hundred others without a name; and mountains, also, whose names, for the most part, are known only to the Indians. The forest looked like a firm grass sward, and the effect of these lakes in its midst has been well compared, by one who has since visited this same spot, to that of a "mirror broken into a thousand fragments, and wildly scattered over the grass, reflecting the full blaze of the sun."

6. Return to Homer's two epic poems *The Iliad* and *The Odyssey* (the Richard Lattimore translations are unsurpassed) and study closely the way in which he uses extended similes taken from the natural world to counter the violence of the human world. Write an essay in which you take a theme from the human world—such as life in the city— and employ similes and/or metaphors taken from the natural world to provide a deliberate counterpoint to the violence, disorder, noise, and confusion.

7. Paint a word picture of an animal in one comprehensive paragraph of at least 150 words in the manner that I did for the river otter (see question 1 in the practice exercises for Chapter 11), using a number of similes and metaphors.

Chapter Eight

Character and Dialogue

The first thing that makes a reader read a book is the characters.
— John Gardner, *The Paris Review*

Dialogue is a very useful tool to reveal things about people.
— Thomas McGuane, *The Paris Review*

Human beings are intensely social animals. We are attracted to works of art not only because we want to understand the universe, but also because we want to better comprehend ourselves. As a result, many of the "classic" works of nature writing are as concerned with human nature as they are with wild nature. For example, readers often find the most appealing parts of Thoreau's *Walden* to be those in which he provides a look at the Irish squatters and freed slaves who lived in "the wilderness" of Walden Pond with him. The same is true of Darwin's "H.M.S. Beagle" journals, where he relates interesting facts and stories about the people of Patagonia or Tahiti. Similarly, Edward Abbey strengthened his book *Desert Solitaire* by focusing as much on the people of Arches National Monument as on the desert uplands and canyons. Peter Matthiessen followed suit in his work *The Snow Leopard*, which was enriched by the inclusion of his traveling companion

George Schaller (perhaps the finest zoologist of the century) and the memory of his departed wife, as symbolized in the elusive snow leopard.

Another major character in these works is the narrator himself. One of the most important aspects of any successful piece of writing is personal revelation—witnessing the author as he or she undergoes a significant change. With Darwin we see an earnest young geologist with an eye for details and a brain that instinctively grasps unifying patterns moving toward his comprehensive theory of natural selection. In the case of *Walden* we observe the process of healing, with the memory of death so deeply submerged in the narrator it is noticeable only in oblique references (such as the black crow motif or the clay-bank passage). *Desert Solitaire* concerns a failed man learning that in solitude and poverty there is companionship, wealth and an unexpected form of success. *The Snow Leopard* tracks the journey of a grieving soul across the desolate landscape of eternity, as symbolized in the Himalayas, and chronicles his discovery there of the austere beauty of faith and hope.

Character and dialogue are so interrelated as to be synonymous. Dialogue is one of the two major means by which character is revealed, in literature as well as in life. The other is action. Both involve decisions. In the case of dialogue, character is revealed by how we inquire, represent ourselves, interact with other people, respond to events and relate to nature. Spoken words are effective tools in terms of character revelation. There is no faster way to reveal character than to let a person speak in his or her idiom through realistic interactive dialogue. The first master of dialogue in American literature was Mark Twain, who was so devoted to the art that he kept detailed notebooks in which he transcribed raw bits of overheard conversation for later use in his stories and essays (John McPhee does the same thing). In this brief excerpt from Twain's paean to the frontier *Roughing It*, we see two American idioms, one representing the country and the other the city, clashing humorously as the semi-literate Nevadan explains to the educated clergymen in town how he wants a funeral to be conducted for his deceased friend:

"Are you the duck that runs the gospel mill next door?"
"Am I the—pardon me, I believe I do not understand?"

With another sigh and a half sob, Scotty rejoined:

"Why you see we are in a bit of trouble, and the boys thought maybe you would give us a lift, if we'd tackle you— that is, if I've got the rights of it and you are the head clerk of the doxology works next door."

"I am the shepherd in charge of the clock whose fold is next door."

"The which?"

"The spiritual adviser of the little company of believers whose sanctuary adjoins these premises."

Scotty scratched his head, reflected a moment, and then said:

"You ruther hold over me, pard. I reckon I can't call that hand. Ante and pass the buck."

"How? I beg pardon, what did I understand you to say."

"Well, you've ruther got the bulge on me. Or maybe we've both got the bulge, somehow. You don't smoke me and I don't smoke you. You see, one of the boys has passed in his checks and we want to give him a good send-off and so the thing I'm on now is to roust out somebody to jerk a little chin music for us and waltz him through the handsome. . . . "

Generally dialogue is used by prose writers in four ways: as an opening or closing device (as seen in Chapters 4 and 5), as a transitional device to link together paragraph blocks, or as a means of expressing the climatic revelation or action. In the case of the excerpt from *Roughing It*, the long run (five pages) of largely unattributed dialogue between "Scotty" Briggs and the clergyman from "an Eastern theological seminary" forms the centerpiece and climax of a chapter meant to underline the differences between east and west, town and country, city-educated and wordly-wise. More often among nature writers, dialogue is used more sparingly than this, with perhaps a few lines in every three or four pages of expository prose. Sometimes it is even less than that. In Isak Dinesen's *Out of Africa*, for example, a powerful nature book inspired by the death of her lover Denys Finch-Hatton, the chapter devoted to Finch-Hatton ("Wings" in

Part IV—"Visitors to the Farm") has only a few odd lines of dialogue
sprinkled here and there like flowers on a grave. Sir Denys arrives back
from safari and asks, "Have you got a story?" (No response in dialogue
from Dinesen). Several paragraphs later, as they sit in her house in the
Ngong Hills he observes that "I would like Beethoven all right, if he were
not vulgar." (More expository prose from Dinesen). Later, when she and
Sir Denys are out hunting they spot a lion and he asks, "Shall I shoot
her?" (Again no response in dialogue from the narrator). And so on for
approximately twenty pages, bearing in mind that this is the climatic essay
in the book. Dinesen uses these single lines of dialogue as pivot points
from which to turn the narrative in new directions, i.e., as transitional
devices. There are no attempts at building any revelatory dialogue
between her and the chief character in the book. Curiously, she concludes
the essay on several lines of dialogue between herself, "a very old Kikuyu"
native, and Sir Denys:

> "You were up very high to-day," he said, "we could not
> see you, only hear the aeroplane sing like a bee."
> I agreed that we had been up high.
> "Did you see God?" he asked.
> "No, Ndwetti," I said, "we did not see God."
> "Aha, then you were not high enough," he said, "but
> now tell me: do you think that you will be able to get up high
> enough to see him."
> "I do not know, Ndwetti," I said.
> "And you, Bedar," he said, turning to Denys, "what do
> you think?
> Will you get up high enough in your aeroplane to see
> God?"
> "Really I do not know," said Denys.
> "Then," said Ndwetti, "I do not know at all why you
> two go on flying."

The last line alludes rather directly and ironically to the fact that Sir Denys
Finch-Hatton was soon killed in the same aircraft.

Sometimes, however, contemporary nature writers do find it necessary to utilize interactive dialogue to advance plot and achieve character revelation in the manner of Twain. A case in point is Terry Tempest William's *Refuge*, a book in which her relationship to her dying mother is as significant as her relationship to the dying Bear River Migratory Bird Refuge. Here Williams skillfully uses dialogue quite frequently to provide the reader with a detailed look at the dynamics of the mother-daughter relationship:

"How much should I tell her?" Dr. Smith asked me in his office. Mother was in the examining room.

"Tell her the truth," I said. "As you have always done."

I could feel the tears well up in my eyes. I was trying to be brave.

"You can't be surprised, Terry. I thought you had accepted this last summer."

"We did. I mean I had, but hope can be more powerful and deceptive than love."

"Her weight loss of eight more pounds is not a result of flu. It's the cancer. She doesn't have much time," he said. He walked out and opened the door to Mother's room. After the examination, he came back out and said things looked better than he thought, that the tumor he had felt in June was gone, and that the others felt smaller. Mother was very quiet. On the way home, I asked, "What do you think?"

"It doesn't really matter, does it?" she said. "Let's just take one day at a time." I had the sense that she wanted to cry. And I thought of her mother, how once in the nursing home, after we had been crying together, I said, "Oh, Grandmother, doesn't it feel good to cry?" and she replied, "Only if you know there is an end to your tears."

There is probably no other way that this incredibly sad scene could be presented. Summation would not convey the emotions as effectively as the spoken words of the chief characters, nor would expository description do the intense feelings justice.

With respect to action, character manifests itself not in spoken words but in what people decide to do—the route to take, the goals to seek, the reactions to adversity, the friendship made, the promises kept. Action appears at similar points in the narrative as dialogue—openings, closings, transitions, and climaxes—but is particularly useful in climaxes. Generally speaking, action is less effective as a revelatory device than dialogue, which puts into precise words what action only shows or implies. One of the best examples of an author using action to provide character revelation is found in Edward Abbey's essay "Havasu," which occurs about two-thirds of the way through *Desert Solitaire* (a naturally climatic point in the overall narrative). Abbey has spent five weeks living by himself at the bottom of the Grand Canyon. What at first seemed an Edenic paradise has turned into a solipsistic hell: "I lived narcotic hours in which like the Taoist Chuang-tse I worried about butterflies and who was dreamer what . . . I went for walks. I went for walks. I went for walks and on one of these, the last I took in Havasu, regained everything that seemed to be ebbing away." What happens on this crucial walk is that Abbey becomes trapped in a side canyon, symbolic of the penultimate trap of solipsism, and fears, through internal monologue, that he will die there: "After the first wave of utter panic had passed I began to try to think. First of all I was not going to die immediately . . . My second thought was to scream for help [but no one would hear] . . . How about a signal fire? There was nothing to burn . . . I began to cry. It was easy. All alone, I didn't have to be brave." Somehow our hero is able to extract himself from this certain death and then spends the night sleeping under a rock ledge: "It was one of the happiest nights of my life." The side-canyon is both literal and figurative; a nameless side-canyon in which the narrator may very well have had some problems, but also a metaphor for the dangers of withdrawing too far into the back canyons of the self. In this way, Abbey has made his point about the necessity of human contact, and has also entertained us with a good action-packed story along the way.

Character becomes the sole subject of nature writers when they write an essay that focuses on a single person as being representative of a particular place or point of view. Charles Bowden did this, for example, in his book *Frog Mountain Blues*, which is devoted to the Santa Catalina Mountains north of Tucson, Arizona. Bowden focuses an entire essay on Buster Bailey, a curmudgeon who lives in the foothills of the Santa Catalina. Buster is presented as a symbol for the indefatigable local people who have become like the tough old mountains in which they struggle to exist. Other authors have done this in varied ways. N. Scott Momaday, for example, focuses on his grandmother, a Kiowa elder, in his book *The Way to Rainy Mountain* as being representative of the region where the tribal reservation was located: "A single knoll rises out of the plain in Oklahoma, north and west of the Wichita Range. For my people, the Kiowas, it is an old landmark, and they gave it the name Rainy Mountain. The hardest weather in the world is there." Her grave is nearby and invests the land with a spirit and the mountain with a weight that neither possessed before her life and death. Similarly, Peter Matthiessen's book *Sand Rivers*, which is devoted to the Selous Game Reserve in Tanzania, takes a long look at the life of Constantine Ionides, the "father of the Selous." Matthiessen and his companion, a refuge manager, pay a reverential visit to Ionides's remote wilderness grave at the base of Nandanga Mountain. Ionides is seen as a man who embodied the rugged individualism that is necessary to survive in the outback of East Africa.

An essay with no people and plenty of expository writing can still be quite powerful, can endure, can meet the expectations of the reader, the reviewer, the critic. Stephen Jay Gould, the popularizer of paleontology, has succeeded in this mode for years while writing his monthly column for *Natural History* magazine. On a similar note, Aldo Leopold's greatest essays—"Thinking Like A Mountain" or "The Land Ethic"—are very dry when compared with the wilderness essays and writings of Bob Marshall, and yet Leopold's are read and discussed more often. All in all, though, the nature essayist is probably improving his or her chances at universalizing theme and extending the life of the work by including dialogue and making some sort of an attempt at character. What many of us have discovered, who have devoted our lives to wild nature, is that what you finally discover out there—as Peter Matthiessen did in the Himalayas, for example—is that you can not escape human nature. The two are inextricably intertwined, and what our readers most hunger for are the connections between the two, the intersections which are revealed most clearly through dialogue and character.

Practice Exercises

1. Use a tape recorder to record a conversation. Afterwards, transcribe the dialogue. What sorts of idioms (colloquial expressions) are in evidence? Are there any patterns or rhythms discernible? How could you cut the transcript and still achieve the same effects? What can you learn about the manner in which people express themselves by paying such close attention to dialogue?

2. Write an essay in which you closely link a particular character to his or her region of the country, as Charles Bowden did with Buster Bailey in *Frog Mountain Blues* or as N. Scott Momaday did with his grandmother in *The Way to Rainy Mountain*. An intrinsic part of that essay could be a dialogue between you and that person, in which you capture something of the essence of the way they represent themselves with the spoken word.

3. In the Mono Lake essay by Mark Twain we saw the clash between two different regions of the country in the dialogue between the miner and the minister. Research and write an essay in which, similarly, you achieve a counterpoint between two vastly different personalities or cultural viewpoints through the use of dialogue. For example, you could ride a ski lift and carefully listen to how various people from different regions and parts of the world relate to the snow, the mountain, and the whole concept of ski resorts and nature. If you live by the sea, you could take an all day fishing trip and, again, pay close attention to how those on board interact with each other and the sea. The point is to bring into close juxtaposition vastly different worlds and see what sorts of humorous interactions arise, with the background foreground, and mid-ground always being, of course, nature.

Chapter Nine

Story-Telling

I am always at a loss to know how much to
believe of my own stories.
— Washington Irving, *Tales of a Traveler*

The Roman poet Quintus Horace, the bard of the Sabine Hills, suggested in *Ars Poetica* that the purpose of literature is "to edify and to entertain." Nature writers have, in this sense, a dual responsibility: to educate their readers about the processes of nature, and to entertain them with a memorable story along the way. Sometimes this means slightly altering character, plot-line, and even setting in order to tell a worthwhile story. Along these lines, the Texas folklorist and writer J. Frank Dobie once said he would "never let the truth stand in the way of telling a good story." It is important that the non-fiction writer understand there are acceptable methods for changing character identity or creating composite characters, for intentionally crossing the fact/fiction boundary, and for rearranging chronology in order to make a more appealing story. It is also essential to recall that essay writing is not journalism. The nature essay, which frequently becomes a sort of non-fiction story, is not meant to be a factual photograph; rather it is reflective and, to use Samuel Johnson's word, "rambling." The personal essay, in fact, is the antithesis of the news-oriented article that is depersonalized and informationally based. The essay is expressive, and, to the extent that all perception involves some personal bias based on perspective and interpretation, imaginary.

In an often-cited passage in his memoir *Oil Notes* Rick Bass, speaking of what he called the "deceit in writing," said that the task of the writer consists of "building a lie and then swinging the lie's massiveness into the path of the reader and hiding behind it." While lecturing at the University of Alaska in 1991, Bass frankly indicated to the audience how several portions of that book, a work of non-fiction, improved on the real events in the interest of greater effectiveness. All non-fiction writers occasionally fictionalize to the extent required by circumstances, whether Washington Irving in 1849 or Rick Bass in 1985. In this sometimes tricky process they are guided by common sense and precedent. For example, when Henry David Thoreau wrote his book *A Week on the Concord and Merrimack Rivers*, he compressed two weeks on the rivers into one week, deleted the week spent off the rivers, and artificially shaped the trip into daily chapters with philosophical themes and frequent poetic asides. He also sandwiched in a large quantity of information from other sources, including popular field guides and regional histories. The effect of all this interpolated material, incidentally, was not as successful as he had hoped. Only 294 copies of the first printing were sold. When the publisher returned the unsold copies to him Thoreau wrote, "I have now a library of nearly nine hundred volumes, over seven hundred of which I wrote myself" (Journal—October 28, 1853).

More recently, Edward Abbey fictionalized large portions of his non-fiction book *Desert Solitaire*: he compressed two seasons in the park into one season; he obviously invented the story of the boy who dies before being able to tell his story in the essay "Rocks"; he took some liberties with the story of the wild horse in "The Moon-Eyed Horse" (this according to his friend David Petersen); he took great liberties with the story in "Dead Man at Grandview Point" (he was not, according to some rangers, even there during the search); he exaggerated or improvised the snow-sliding anecdote in "Tukuhnikivats" (this again according to Petersen); and, most interestingly, he completely wrote his family out of the book (they stayed with him at Arches for at least part of one summer). Those rangers who have since published their accounts of the summers Abbey spent in Arches, contrasting the "facts" of their recollections with the creative story told in *Desert Solitaire*, evidently do not appreciate the difference between journalism and essay writing. Abbey was not reporting on Arches National Monument (now a national park) for *National Geographic* magazine; rather, he was relating the story of an idealized season in a very special place with all of the tools traditionally used by essayists: he fictionalized for dramatic effect or to emphasize key points; he created composite characters or changed identities to more effectively advance narrative and to protect people's privacy; he removed what was not essential to theme, character, and plot; and he altered chronology in order to achieve structural unity. *Desert Solitaire* will never be found in the fiction section of the bookstore or library, but the book, in a very fundamental sense, meets the criteria by which both are defined.

It's difficult to provide novice writers with specific guidelines on these issues, because none exist. Every topical situation the writer encounters is different and any attempt at formalizing rules would be at best reductive. Probably the best guidance can be found in examples. Let me offer one from my own writing. Several years ago I wrote an essay entitled "The Coral Reef at Akumal" for my Caribbean book *The Islands and the Sea*. This essay relates the experience of swimming out two hundred yards across a coral lagoon to the barrier reef that parallels the Yucatan Peninsula of Mexico. The purpose of the essay is to provide the reader—particularly the reader who will never put on a mask and flippers and see a coral reef firsthand—with a vivid sense of that incredible experience. I stayed at the village of Akumal for three days; on the first two days I swam to the reef with my wife and another couple staying in the next beach hut. On the last day, before we left for Playa Del Carmen and the ferry to Cozumel Island, I swam out to the reef by myself. The essay, following the precedent established by such writers as Thoreau and Abbey, undertakes the following: it compresses the highlights of three trips into one; it eliminates the other people because that would distract the reader from the primary focus of the essay; and it adds some scientific information that I learned from discussions with specialists upon my return to the United States (such as information on the life-cycle of coral). If I had had more room (my essay was limited to 7,500 words), I would certainly have enjoyed describing my lively companions and also the wonderful little natural history museum to be found in one of the huts (a sort of pre-Linnean "cabinet of curiosities"), but space precluded that. Given the short format, I could only concentrate on one unifying action and a single theme, with the result being an altered chronology and crystallized experience.

One of the earliest reviews of Norman MacLean's fictionalized memoir, *A River Runs Through It,* compared the story to Thoreau's first book, *A Week on the Concord and Merrimack Rivers.* Walter Hesford, a professor at the University of Idaho, pointed out that both books, in addition to being concerned with fishing, were elegies for a brother lost to early death. In both cases the authors had been left shattered and achieved a catharsis through writing. MacLean is useful to examine on this issue of story-telling because he was a distinguished English professor at the University of Chicago and an accomplished story-teller and writer. He wrote one work of fiction, *A River Runs Through It,* and one work of non-fiction, *Young Men and Fire.* In the first instance, his primary problem was that the story of his life in Montana was very long—crossing decades—and structurally complicated. MacLean had to reorganize some of the action and dispense with some other details and minor characters. Most importantly, he had to change the location of his brother Paul MacLean's murder in 1938 from Chicago to Missoula, in the interest of achieving a more emotionally charged climax. MacLean unified the whole, in his own words, by showing "the order in which you learn about the art of fly-fishing, from the initial scene in which my brother and I [are boys] . . . to the later scene where you saw his "brother, the master fisherman . . . landing his last big fish." So the story he created, though essentially a personal memoir, was finally submitted to publishers, accepted, and sold as a novella. Other authors, like Abbey in *Desert Solitaire,* have changed more and called it non-fiction.

Young Men and Fire, on the other hand, concerns the 1949 Mann Gulch fire in Montana and the thirteen smokejumpers who died there, and is purely a work of non-fiction. The book, a work of reportage, is based on an extensive period of research that involved reviewing the U.S. Forest Service file sources, interviewing the survivors and other participants, and physically walking the dry ridge where the men perished. *Young Men and*

Fire is also an oblique elegy to his wife Jessie MacLean, who contracted emphysema and met a painful asphyxiative death similar to that suffered by the thirteen smokejumpers before the fire reached them. Jessie MacLean's death is mentioned in the closing as a sort of framing device:

> I, an old man, have written this fire report. Among other things, it was important to me, as an exercise for old age, to enlarge my knowledge and spirit so I could accompany young men whose lives I might have lived on their way to death. I have climbed where they climbed, and in my time I have fought fire and inquired into its nature . . . Perhaps it is not odd, at the end of this tragedy where nothing much was left of the elite who came from the sky but courage struggling for oxygen, that I have often found myself thinking of my wife on her brave and lonely way to death.

The two stories that MacLean committed to publication, then, were, like so many works of nature writing (*Walden*, *Ranch Life and the Hunting Trail*, *Out of Africa*, *The Solace of Open Spaces*, *The Snow Leopard*, *Land Circle*, *Refuge*) inspired by the loss of a loved one. MacLean chose different genres because the stories demanded different forms, but the styles—lean, muscular prose stripped of modifiers and intensifiers—remained the same, as did the biblical theme: the injustice of death.

It is important to distinguish between an anecdote and a story. An anecdote relates an incident interesting enough to be memorialized. It has a beginning, middle, and end, but lacks the depth of feeling, the personal revelation, the universality, stylistic innovation and sophistication of craft that is present in a serious story or essay. Let me give an example of both an anecdote and a story so that the difference is clear. In his essay "The Stricken River" (from *Travels in Alaska*) John Muir relates, among other things, the account of how he and a missionary named Young undertook a hike of around fourteen miles near Glenora in coastal British Columbia. Just as they were nearing their objective Young fell down in a rock field and dislocated both of his arms. Muir was able to relocate one of the arms. The other could not be set. Somehow the two made it back to the bottom of the valley where the boat was waiting for them. After consuming some brandy, Mr. Young lay down and had his other arm put back into place. Muir writes that the only reason he committed the tale to print is that "a miserable, sensational caricature of the story . . . appeared in a respectable magazine [back East]" and he "thought it but fair to my brave companion that it should be told just as it happened." Here the author has made it clear that this is not a story in the literary sense, but is, rather, a quickly sketched out anecdote offered to brace the reputation of a friend. Nature writers are often called upon to include such anecdotes in their essays , and should not feel compelled to approach them in the same comprehensive way they do stories such as those we saw with Abbey and MacLean.

From the first paragraph of Alaskan poet laureate John Haines's essay "The Sack of Bones," the reader knows that this is to be a genuine story, full of resonance, implications and artistry, and not just a curious tale:

> I heard this story from Hans Seppala late one summer evening. We were sitting in his cabin at Shaw Creek . . . Out the open door of the cabin, in the midnight dusk, we could hear the creek flowing by, but with hardly a sound in its slow, brown current. The landscape held that unusual quiet, when for an hour or so, before the sun lights up the hills again, the life of the arctic summer is stilled, and few birds sang. Hans had his fund of stories, which he told with particular emphasis in his own kind of English, generous with obscenities, and half-formed on the syntax of his native Finn . . . he told many of the same stories over and over, hardly changing the details . . . Most of his stories were about people we both knew . . . But this story was different; he told it once, and I never heard it again.

The opening paragraph—with its "midnight dusk" and "slow brown" stream and "unusual quiet"—echoes the beginning of Joseph Conrad's *Heart of Darkness*, where the men sit on a boat along the Thames as the tide turns and Marlowe begins his saga. The second paragraph builds on this expectancy, as the word story is repeated four times for emphasis. Also, we learn that the narrator, Hans Seppala, has only told the story once and that Haines "never heard it again." Haines has skillfully hooked the reader, who reads on with the sense that this is a singular account of some weight and meaning.

What follows is a Jack London-esque drama of greed, murder and cover-up set in the frosty north. Three men are involved. Two of them—Fred Campbell and Emory Herschberger—trap a place called Shawk Creek. A third, known only as Martin, trespasses. A dispute ensues and Martin disappears. Two years pass. Hans Seppala is driving his sled team along the Tanana River and comes upon a pile driftwood. Inside is a canvas sack secured with heavy wire. Seppala looks closer and sees the bones are small and light. The bones of a human being. Three weeks pass before Sepalla can return and when he does the sack of bones has vanished in the runoff of breakup. Everyone suspects Fred and Emory "but no one would ever know for sure . . . and the Tanana kept its secrets." The story closes with Seppala and Haines back at the cabin:

> We sat there thinking about this strange event, the coffee gone cold in our cups. Morning brightened in the forest beyond Hans's clearing, a fine mist came off the water of Shaw Creek . . . He turned, and looked at me sharply and strangely through the steel rims of his spectacles.

Here Haines has given the reader far more than a mere anecdote. The story is invested with meaning and a moral and through it all nature—the relentless uncaring Tanana—is as strong a character as Seppala or Martin or Haines himself. These are some of the differences between an anecdote and a story.

Norman MacLean once wrote that he had learned "a great deal about the sound of literature" by "listening to story-tellers around the camp-fire at night." Much of the power of MacLean's stories derives from the fact that he was keenly aware that all written literature is based in oral literature; he understood that a story must first "sound" good before it does anything else. Another comment of his on this subject is worth recalling:

I write paragraphs which I hear as units of sound that rise a few notes to their middle, then drop a note or two as the paragraph comes to an end—except right at the very end when they rise a note or two to start the next paragraph, slightly higher in the scale than the preceding paragraph, and so help quietly to increase the scale of interest as the story proceeds.

By this description, MacLean wrought each paragraph as carefully as if it were its own self-contained story. Such attention to details is the definition of excellence and one of the guarantees of success in narration, whether written or oral.

For each of the past six summers I have spent two weeks in Denali National Park photographing wildlife. The park road, a pretty decent gravel road, runs about ninety miles from Riley Creek on the east to Wonder Lake on the west. Quite often, I have people accompany me on these trips, either friends or park visitors that I meet in the Teklanika campground where I make my headquarters. Much of the time, as we are driving slowly in search of grizzly bears and caribou, not much happens. On some days, we may drive for ten or twelve hours and see very little wildlife, perhaps a single grizzly bear digging for peavine on a distant gravelbar or a band of Dall sheep sleeping in the shadow of a crag. To pass the time I tell stories. I tell them to different people, and note their reactions. I learn what works and what does not work, what is interesting to them and what is not, what should be emphasized and what should be forgotten. After perhaps three or four tellings, I finally write the story down. I think it is a good idea to tell a story in this way, with the audience right beside you. If you can keep a person interested and entertained on a hot dusty road in the middle of nowhere, then chances are your story might have a chance of surviving on the windy plains of time.

Practice Exercises

1. Write out, word for word, an essay you admire that is concerned primarily with telling a good story. How has the writer planted the seed of interest? How has he or she sustained that interest? Where is the climax? How does the writer make an exit after having reached that point? Virtually anything written by Mark Twain or Rick Bass (*The Deer Pasture*, *Wild to the Heart*) would be worth utilizing in this regard. Both are consummate storytellers.

2. Pick an interesting nature story that has been in your local news recently and write the narrative out in essay form. For example, you could write about the beaching of a group of whales (as Barry Lopez did in his essay "A Presentation of Whales" in *Crossing Open Ground*). The important thing is to tell a story about something that you passionately care about, to convey the immediacy of a lived or cherished event, to universalize a particular experience.

3. One of the questions of chief interest in this chapter was the extent to which essayists fictionalize real events. Write about an experience that was personally experienced; as you do pay close attention to where you find it necessary to fictionalize in the interest of achieving structural unity, dramatic intensity, or character revelation.

4. Interview several observers of or participants in an event, and then write an account in which you focus on how perspective and self-interest biases perception. I had an experience with this phenomenon myself several years ago, following an incident in Denali National Park in which eleven wolves attacked and killed two grizzly bears. When I interviewed several eyewitnesses, and later reviewed the summary account prepared by the National Park Service, I was astonished at how varied the stories were. What made it all the more surprising was that these were trained biologists, and yet, at several points in their respective narratives, it seemed as though they had observed a totally different event. If you live in south Florida, you could write

about panther sightings. If you're from Tennessee, you could write about eastern mountain lion sightings. Similarly, in Colorado there have been many alleged grizzly bear and wolf sightings over the years. Any of these topics would make for an interesting story about the nature of human nature as well as wild nature.

Chapter Ten

Style

When we come across a natural style, we are surprised
and delighted, for we expected an author, and we find a man.
—Blaise Pascal, *Pensées*

Thou hast most traitorously corrupted the youth of the realm
in erecting a grammar school . . . It will be proved to thy face that
thou hast men about thee that usually talk of a noun and a verb, and
such abominable words as no Christian ear can endure to hear.
—William Shakespeare
Henry IV, Part II (Act 4, Scene 7)

A good writer can craft clear, concise and cohesive prose that also reflects
his or her personality. When this occurs, we say that the writer has developed and mastered his or her own personal idiom, and has created a personal style. The word style derives from the Latin *stilus*, which was the
sharp piece of bone with which Roman writers inscribed their wax note
tablets. In fact, almost all writing styles you will encounter in nature writing, whether the informal colloquial style of Edward Abbey or the formal

elegant style of Ralph Waldo Emerson, derive from Latin sources, as we shall see more fully later in this chapter. Style is most fundamentally concerned with sentences and paragraphs. These are the building blocks by which writers construct arguments, tell stories, create word pictures, organize thoughts, and bring people to life on the page. Sentences are divided into those that are simple, that have only one subject and predicate, and those that are complex that have more than one subject and predicate. Edward Abbey, for example, preferred the former and Emerson worked frequently with the latter and this tells us a lot about their respective styles. Similarly, the paragraphs used by writers vary considerably in size and shape, ranging from those based on a moderate number of sentences of fairly simple construction, as with Abbey, to lengthy paragraphs composed primarily of complex periodic sentences, as are found with Emerson.

Style is usually acquired over time, as the author studies and practices various paradigms and approaches and gradually synthesizes and transcends influences to create a personal idiom. There are, however, always exceptions like Hemingway, who at a very young age produced a wholly unique style (Stephen Crane, Ambrose Bierce, Jack London as precedents), and whose style then went on to become the most influential in the century (with credit given to him by writers as diverse as Camus of France, Kawabata of Japan, and Solzhenitsyn of Russia). Generally speaking, the greatest writers are those whose style can accommodate the most comprehensive tonal range. Writers who have accomplished this in the twentieth century include William Faulkner (whose style could be used for an old-fashioned hunting story like *The Bear*, a hilarious picaresque like *The Rievers*, or a dark tragedy like *The Sound and the Fury*) or Gabriel Marquez (whose style can range from a hilarious satire like *A Very Old Man with Enormous Wings* to a poignant love story like *Love in the Time of Cholera*). Writers who have mastered their individual style, and whose innovative style then goes on to influence their peers and younger generations of writers, often have that achievement recognized in the granting of major literary awards, as occurred with each writer mentioned in this paragraph when he was summoned to Stockholm.

The formal style, referred to earlier with respect to Emerson, can be broken into three categories. These range from writing like Emerson's, which is based on refined latinate diction and vast elegant sentences and paragraphs, to writing more like Terry Tempest Williams's, which is nearly scriptural in its simplicity and majesty and utilizes primarily the oldest Anglo-Saxon words to create an unadorned prose surface. In between is a sort of "middle style" sharing attributes of both. Let's first look at two representative paragraphs from Emerson and Williams to refresh our memories:

> The ancient Greeks called the world κοσμοζ [cosmos], beauty. Such is the constitution all things, or such the plastic power of the human eye, that the primary forms, as the sky, the mountain, the tree, the animal, give us a delight in and for themselves; a pleasure arising from outline, color, motion, and grouping. This seems partly owing to the eye itself. The eye is the best of artists. By the mutual [sic] action of its structure and of the laws of light, perspective is produced, which integrates every mass of objects, of what character soever, into a well colored and shaded globe, so that the particular objects are mean and unaffecting, the landscape which they compose, is round and symmetrical . . . [and so on]
> —from Ralph W. Emerson's essay "Nature" (1836)

> It is snowing at Bear River in May. I can only drive out three miles west of Brigham City. The lake stops me. Before the flood, it was a fifteen mile trip. The waves of Great Sale Lake are lapping just below where my car door opens. Gray sky. Gray water. I have the sense that I am suspended in the middle of the lake with pelicans, coots, and grebes. I keep driving with the illusion that my old Peugeot station wagon is really a boat. When the lake starts seeping into the floorboards, I come to my senses. I stop the car, carefully open the door and climb on the roof.
> — from Terry Tempest Williams's essay "Long-Billed Curlews" in *Refuge* (1992)

In the first instance, we have complex sentences with multiple clauses (subject/predicate pairs) and sentence lengths that range from forty-six to fifty-one words. In the second case, we have simple sentences in all but one case, and the longest sentence consists of only nineteen words. Williams also uses a more Anglo-Saxon diction that produces her desired effect of restrained emotion, whereas Emerson uses the more varied diction and modifying patterns of the parlor room philosopher and Lyceum circuit lecturer. Both styles are popular with nature writers in modern times, with the elegant style being practiced by such writers as Theodore Roosevelt, Isak Dinesen, Rachel Carson, and Barry Lopez, and the laconic style being found in writers such as Peter Matthiessen, Dan O'Brien, Harry Middleton, and John Haines.

Between the elegant style of Emerson and the laconic style of Williams is the middle style, which is more flexible and can accommodate some of the characteristics of either extreme; writers like Mark Twain have preferred this middle ground:

> The first time I ever saw General Grant was in the fall or winter of 1866 at one of the receptions at Washington, when he was general of the army. I merely saw and shook hands with him along with the crowd but had no conversation. It was there also that I first saw General Sheridan. I next saw General Grant during his first term as President. Senator Bill Stewart of Nevada proposed to take me in and see the President. We found him in his working costume, with an old, short, linen duster on it and it was well spattered with ink. I had acquired some trifle of notoriety through some letters which I had written in the New York Tribune during my trip round about the world in the 'Quaker City' expedition [Twain's book *Innocents Abroad* (1869)]. I shook hands and then there was a pause and silence. I couldn't think of anything to say. So I merely looked into the General's grim, immovable countenance a moment or two in silence and then I said: "Mr. President, I am embarrassed. Are you?" He smiled a smile which would have done no discredit to a cast-iron

image and I got away under the smoke of my volley. I did not see him again for some ten years. In the meantime I had become very thoroughly notorious . . . Carter Harrison, the Mayor of Chicago . . . walked over with me [to President Grant at the Parker Hotel in Chicago] and said, "General, let me introduce Mr. Clemens." We then shook hands. There was the usual momentary pause and then the General said: "I am not embarrassed. Are you?"

—from Mark Twain's autobiography (1917)

Twain used the middle style because it fit his humorous treatment of subjects and his vast range of topics well—he could easily slip in colloquialisms, bits of dialogue, asides and digressions without losing the comprehensiveness afforded by the larger paragraphs and paragraph blocks. He also could vary his sentences, from short and staccato to long and ponderous, to replicate the patterns of spoken English. More recent practitioners of the middle style include John Muir, Norman MacLean, and David Rains Wallace.

The informal style is largely vernacular and conversational, but can move towards formality at times, usually in the closing paragraphs. The best-known exemplar of this style in Emerson's age was Walt Whitman and the most representative in recent times was Edward Abbey. You find the informal style throughout Abbey's essays and in several of his novels, especially his last serious novel *The Fool's Progress*. The informal style of writing is distinguished by its use of slang and even profanity, its odd right turns and digressions, its frequent ungrammatical constructions, and its use of the second and first person. The tone is approximately that of a personal friendly letter from the writer to the reader. No experience or bodily function is too private to be withheld, no literary risk too great to be taken, no topic or opinion is off limits. Edward Abbey's essay "A Desert Isle" exemplifies the informal style. A typical passage will illustrate:

I found a small sandy beach in a sheltered cove, took a swim, stretched out on the warm sand, hat over my face, and let the sun blaze down on my body. The sun, the sand, the clamoring sea, my naked skin. Close to peace for the first time in weeks, I began to think of women. Of this one, that one, all the lovely girls I've found and known and lost and hope to find again. That girl in Tucson, for example: her light brown hair, her docile eyes, the glow of her healthy flesh.

"Take me to Mexico," she had said.

"I'd rather take you tonight," I said.

God damn it. Really a mistake to come to a perfect place like la Sombra without a good woman. God damn it all. I committed adultery with my fist and went back to see how Clair was doing.

Still later the prose assumes a more formal tone:

Death Valley by the sea. Salmon-colored clouds float over the water. Reflecting that light, those images, the sea, now still, looks like molten copper. The iron, wrinkled, savage mountains take on, briefly, a soft and beguiling radiance, as if illuminated from within. Canyons we have yet to look at—deep, narrow, blue black with shadow—wind into the rocky depths. Sitting on a hill above our camp, listening to the doves calling far our there, I feel again the old sick romantic urge to fade away into those mountains, to disappear, to merge and meld with the ultimate, the unnamable, the bedrock of being. Face to face with the absolute whatever it is. Sweet oblivion, final revelation. Easy now. What's the hurry? I light a cigar instead.

Characteristically, Abbey switches back to the personal voice in the final three sentences. The informal style is practiced, either regularly or occasionally, by a significant number of contemporary nature writers: Rick Bass, Jim Harrison, Bill Kittredge, John Gierarch, Ed Engle, James Crumley, Russell Chatham, Tom McGuane, Bob Shacochis, Doug Peacock.

A word or two about origins, which might help writers understand where their styles have evolved from. The elegant style of Emerson derives from the writings of Cicero (106–43 BC), who was a Roman attorney and philosopher. Cicero was accustomed to speaking expansively in the manner of lawyers before skeptical juries (as he did most brilliantly in winning the impeachment of Verres for corruption in 70 BC). Cicero published numerous essays and speeches, and, though not read much by nature writers today, was closely studied by those in the nineteenth century, such as Emerson and Thoreau, whose writings now influence contemporary essayists. The laconic style practiced by Terry Tempest Williams comes to us from the writings of Caesar (100–44 BC), a career military officer who later entered political life. Caesar was accustomed to the brevity and urgency of command, to clear orders given quickly with short declarative sentences. He operated for many years in the wild fields and forests of France and Britain, his life in constant danger as he fought hostile armies far from the comforts of Rome. Caesar preferred the simplest and most direct statements, anticipating Hemingway by 2000 years with sentences like "There is nothing, absolutely nothing, fairer, more beautiful, more to be loved, than courage." Caesar's *Gallic Wars* has been the text by which beginning Latin students have learned the language for a thousand years (who knows but that a thousand years from now the dead language of American English will be learned by students reading U.S. Grant's *Memoirs*, which is also a masterpiece of concision and felicity and which was edited by Mark Twain). The middle style derives from a number of sources, most notably Cicero's letters, which lack the formality of his published writing and powerfully influenced later writers of the essay including Montaigne; and Erasmus, a Renaissance writer whose writings energetically praised the vernacular and the commonplace in the spirit of those revolutionary times. Lastly, the informal style comes out of a line that goes back to such writers as the Roman comic playwrights Plautus and Terence, who incorporated the vernacular, colloquial idioms into Roman literature in the same way that Aristophanes secularized Greek literature.

Although I have experimented with the variations discussed above over the years, I have always written for publication in the middle style. For me it has been a question of comfort. The informal and two variations of the formal mode have never "felt" right, for lack of a better word. It is important to recall for readers that there is no right or wrong style. A style should be chosen that fits your personality, your manner of thinking, and your way of expressing yourself. One could hardly imagine a worse hell than to try to teach Edward Abbey to write like Ralph Waldo Emerson, or vice versa. What appeals to me about the middle style is its diversity, the almost endless possibilities for topics and treatments that the other options do not always or easily afford. The middle style here permitted me to use sentence fragments where necessary, to employ extended paragraphs or shorter paragraphs as needs dictate, to make casual personal references; and to range from humorous asides to more serious factual reportage. For me, the middle style represents that broad continuum between being overly formal and solemn and overly frank and informal.

It is customary for the writer on style to provide a list of recommendations as part of his or her treatment of the subject, as Professor Strunk did in his famous book on the subject in 1919 (when the modern world and the modern style was born). Here are a few—let us call them reminders—for your consideration, in no particular order. Some of these go beyond style to speak to the process of writing itself:

1. Assume that the reader can draw his or her own conclusions

2. At some point in composition, develop an outline or plan

3. Write in whatever form you are comfortable

4. Aim for concision and clarity

5. Use words precisely

6. Remember that the last word of a sentence, paragraph, or essay occupies a naturally emphatic position, and that what you write here will long echo in the reader's mind

7. Put actions into verbs and actors into the subjects

8. Avoid too many modifiers and avoid intensifiers

9. Write for the ages

10. Describe people and places and things when they first appear in the text, or shortly after

11. Use figurative language judiciously

12. Vary your sentence patterns

13. Put aside your essay for a few days before finishing it

14. Whichever style you choose, master it

15.   Read the best literary examples you can

16.   Write from experience

17.   Do not share your work too soon with others

18.   Establish a deadline as early as you can and keep to it

19.   All grammatical and composition rules can and should be broken where circumstances dictate

20.   Watch your transitions carefully, for it is there that a piece can most easily be lose its connectiveness

21.   Defend the artistic integrity of your work as you would the life or reputation of your child

22.   Humble yourself before the text; put what is best for it above what you may personally care for in it

Style goes back to the roots of literature and language. Style touches on class, as well, in the sense that the formal style has traditionally been the form in which the upper classes expressed themselves, the middle style has appealed to writers from the middle classes, and the informal style has belonged pretty much to the working classes. Style also says something about geography and culture, with the laconic style probably coming to us from the Greeks, from southwestern Asia and also from ancient Egypt, while the elaborate style is possibly more recent and more European in its origins. Style is never constant. It is always changing. What is fashionable today will likely not be as fashionable a century from now. Charles Dicken's monolithic Victorian novel *Bleak House* would not sell as briskly in 1995 as it did in 1853. The best writing, though, transcends style to endure through the ages, like Cicero's letters, Caesar's *Gallic Wars*, or Pliny the Elder's natural history essays. That, it seems to me, should be the aim of style. Lucidity. Forcefulness. Honesty. A sort of lens through which another human being can view the thoughts of another human being, the experience being made more powerful by its being more clear.

Practice Exercises

1. Write a paragraph or two in the informal style of Edward Abbey, Rick Bass, and Jim Harrison. Study their books for clues as to how these authors make that style work—the fidelity to idiom, the fast-paced narrative and quick change-ups, the humorous tone. The advantage of this style is that one can be, and always should be, wholly human. Anything less is a sort of treason. Even if this style is not to your ultimate liking, the exercise will give you a more comprehensive sense of the limits and possibilities of your own style.

2. Similarly, compose a paragraph block in the more formal style of Ralph W. Emerson, John Burroughs, and Mary Austin. Aim to reproduce the complex sentence patterns and longer paragraphs that create the more refined tone of this style. Although it is more difficult in the formal mode to achieve the personal voice so much liked by modern readers, the advantages are certainly there for serious themes, of which there are no shortage (acid rain, nuclear waste, the loss of ozone, global warming, the population crisis). You may find this style ideal for your personality, or you may not. Either way, the practice of an alien style will at least more clearly define your own.

3. The best example of the middle style always has been, and probably always will be, Mark Twain (born Samuel Clemens). After reading a few of his classic "essays" in such works as *Innocents Abroad*, *Roughing It*, and *Life on the Mississippi*, write an essay in which you aspire to the same range and depth, with a bit of humor "sprinkled" here and there as seasoning. Don't, of course, write of his antiquated themes (riverboats, stagecoaches, and the like), but transfer his mode of expression to the modern world, as in a Yosemite Valley tour bus, or a string of Yellowstone dude horses and their cargo, or the lively participants in a Sierra Club knapsack hike through the high country. In each case, their are possibilities for both high and low drama, for serious digressions and comic excursions, and the middle style provides ample possibilities for all.

Chapter Eleven

Fiction and Poetry

I have a genuine love of nature. It is not the least bit affected, but an integral and powerful part of my life. I know that Cooper is a fraud—that he doesn't give a true sense of the sublimity of American scenery. I know that Muir and Thoreau and Burroughs speak the truth . . . Ever since I could walk, I have spent as much time as I could in the open.

> — Theodore Roethke, from *On The Poet and his Craft,*
> *Selected Prose of Theodore Roosevelt*

Even my novels are sometimes classified as nature writing! . . . I never wanted to be anything but a writer, period. An author. A creator of fictions and essays, sometimes poems.

> — Edward Abbey, from the preface
> to the twentieth anniversary edition of *Desert Solitaire*

It is important that naturalists and other writers adopting nature as a theme understand that nonfiction provides only one of several outlets, and that many of our best nature writers have composed regularly in the genres of fiction and poetry, as well. Thoreau, for example, considered himself a poet for years before realizing that prose was his metier (several of these poems are included in *A Week on the Concord and Merrimack Rivers*). Anyone studying Thoreau's poetry is struck by the way in which he learned the value of irony, oblique reference, symbol, image, metaphor, rhythm, internal rhyme, assonance, and alliteration during his long apprenticeship to verse. Similarly, Edward Abbey produced a considerable body of poetry, which has only recently been published, and also wrote a number of acclaimed novels, including two that were made into successful motion pictures ("Lonely Are The Brave" and "Fire on the Mountain"). Abbey's use of imagery, assonance and alliteration—apparent throughout his most powerful and lyrical work *Desert Solitaire*—was learned in the genre of poetry. Other contemporary writers who work with nature in several genres include Gary Snyder, W.S. Merwin, Linda Hogan, Wendell Berry, John Haines, David Rains Wallace, Leslie Marmon Silko, and Peter Matthiessen.

One of the most interesting nature books written by a novelist in recent times was John Fowles's *The Tree*. Fowles is best known for his novels, which include *The French Lieutenant's Woman*, *Daniel Martin*, and *The Magus*. In *The Tree*, which is in the form of an extended essay, Fowles discusses the importance of nature in his spiritual life and in his works of the imagination. Like the poet Theodore Roethke in the chapter epigraph above, Fowles claims a particularly strong affinity for nature: "I spent all my younger life as a more or less orthodox amateur naturalist... Again and again in recent years I have told visiting literary academics that the key to my fiction, for what it is worth, lies in my relationship with nature... what I gain most from nature is beyond words." With respect to his particular works, Fowles is even more emphatic in showing the influence of nature. For example:

> All through history trees have provided sanctuary and refuge for both the justly and unjustly persecuted and hunted. In the wood I know best [Fowles lives in Lyme Green, on the southern coast of Britain] there is a dell, among beeches, at the foot of a chalk cliff. Not a person a month goes there now, since it is well away from any path. But three centuries ago it was crowded every Sunday, for it is where the Independents came,

from miles around along the border of Devon and Dorset, to hold their forbidden services. There are freedom in woods that our ancestors perhaps realized more fully than we do. I used this wood, and even this one particular dell, in *The French Lieutenant's Woman*, for scenes that it seemed to me, in a story of self-liberation, could have no other setting.

Here we see quite clearly how nature influences, indeed is closely represented, in a work of fiction for the explicit purpose of revealing character and advancing plot. And Fowles does not consider himself alone in having recognized or paid homage to this nexus. "It is not for nothing," he observes, "that the ancestors of the modern novel that began to appear in the early Middle Ages so frequently had the forest for setting and the quest for central theme."

In the most luminous passage in the book, Fowles writes that he sees "trees, the wood, as the best analogue of prose fiction [because] all novels are . . . exercises in attaining freedom—even when, at an extreme, they deny the possibility of its existence." He sees the forest as representing the "'wild,' or ordinarily repressed and socially hidden self" and the novelist's treatment of that through setting and theme as a "return to the green chaos, the deep forest and refuge of the unconscious." This use of nature thematically as a symbol for the mind, for the subconscious, is also apparent in Hemingway's novel *The Old Man and the Sea,* which was the work cited by the Nobel committee in granting him that award in 1954. Like Fowles, Hemingway saw the novel, the artistic process, as a means to transcend self, to become liberated from the constraints of mortality. Hemingway, like Fowles, wisely knew that often the best fictional works are those drawn directly from personal experience. Whereas Fowles placed his novels in the country around Lyme Bay, Hemingway chose the Gulf Stream for the setting of several of his works. The allegory at work in *The Old Man and the Sea,* or at least one of those most popularly through to be at work, is that the sea represents the unconscious, the Marlin represents the work of art extracted from the depths of the unconscious in an epic struggle, and the destruction of the Marlin by the sharks represents what the world does to its creations. In both cases, the novelists have drawn upon a wild nature that they know intimately well in order to provide the setting for their narratives dramatizing conflicts of human nature.

Not all works categorized as being nonfiction are that, exclusively or even at all. In fact, the boundaries between fact and fiction are considerably more blurred than is commonly believed. Quite often nature writers absorb and synthesize what is known about a particular species or subject in order to create an imaginary sequence of events that could have occurred. This is not nonfiction, strictly, nor is it fiction in the classic sense, and yet the books are regularly placed in the nonfiction section of the bookstore and library. Some have called such writing 'faction,' a term that may obscure more than it reveals. A number of years ago David Rains Wallace, a California nature writer, wrote a wonderful book entitled *The Dark Range*, which is a sort of natural history of the night. His setting was the Yolla Bolly Mountains near Sacramento. Each essay examines a particular animal or place in great detail. For example, he devotes part of one chapter to chronicling the nocturnal activities of a black bear. Wallace writes with intimate knowledge of this bear and its perambulations, but obviously—not having radio-tracked the animal—is writing from the imagination, based on facts known generally from primary and secondary sources and personal observations. For example:

> He was about half bear-size and thin enough to be a dog. He was small because he was half grown and thin because his mother had recently disowned him, and he wasn't yet adept at getting food. The peculiar thing about this bear was his fur. On his head and legs it was typically black and glossy, but it grew from his back and sides in long mottled patches of reddish blonde, as if he had been doused with a mixture of peroxide and hair grower . . . Since his mother had driven him off, he had wandered around more or less at random. Nobody had told him that black bears stay in the forest. The previous night he had followed a canyon so far that, looking up from the ground on which he had been tracing an intriguing sequence of smells, he noticed that there were no trees at all on the slopes above. He had been in open places before

this, but the shaggy shapes of pine or fir had always been nearby ... Then events distracted him. A porcupine appeared, another straggler from the woods . . . [the bear's nighttime adventures are then pursued for many more pages]

Wallace is clearly using artistic license here very legitimately to produce a finely wrought picture of place. It is almost as if the text is the voice of the Yolla Bolly Mountains; the narrator has disappeared and the animals, the rocks, the plants, and even the stars speak for themselves. A taxonomic purist would not place a nonfiction stamp on this book, but that is really irrelevant. What is essential is to remind readers, who may be just learning about nature writing in this book, that what Wallace has done here— to fictionalize based on facts—is not only permissible, it is a form of art and is something to be studied and valued.

Poets and nature have had a love affair from the beginning. In Chapter 7 ("Figurative Language") we saw how intimately Homer wove images of rural nature into his war narrative to counterbalance the violence and destruction of battle. Even earlier, around 1000 BC, Solomon, the King of Israel, composed the Songs of Solomon, Ecclesiastes, and The Book of Proverbs, all of which celebrate the beauty of nature and all of which were later incorporated into the Old Testament. Later, the Roman poet Virgil celebrated rural nature in his *Eclogues* and *Georgics*. There was a schism between nature and culture in Western Europe through the Middle Ages, primarily because of the influence of Old Testament decrees of dominion over nature, but the poetry of the Renaissance, culminating in the plays of Shakespeare, was a massive celebration of the beauty and power of nature, as was the poetry of the Romantic period, especially the poems of Wordsworth, Coleridge, and Keats. In the first half of the twentieth century poets like William Butler Yeats, Robert Frost, and Dylan Thomas consistently focused on nature in their poetry. Today, a wide range of poets—W. S. Merwin, Wendell Berry, John Haines—are intimately concerned with nature. Poems offer significant advantages in dealing with nature because they enable a writer to focus almost exclusively on single strong images and to then fuse these images with the music of language into a unified lyric structure. The scope and sweep of prose are not as conducive to producing these solitary moments that can stand alone—poems like Wang Wei's "Wang River Sequence" or Matsuo Basho's "Haiku at the Kashima Shrine."

Most nature poetry is lyric poetry, with the root word coming from the Greek λνρα, or lyre, a stringed musical instrument. In Chinese the root word for lyric poetry is "shih," meaning word song. Although lyric poetry was originally meant to be a song lyric, to accompany a melody played by a musical instrument, this is not usually the case today. Although some lyric poets, like Robert Burns, Robert Johnson, Woody Guthrie, Bob Dylan, or John Lennon have composed their lyrics to accompany a melody and chord progression, many others, like William Wordsworth, William Butler Yeats, and John Haines, have not. The essential thing is to have the sound and sense, the cadence and rhythm, the rhymes and resonance of music replicated in the lyric poem, such as in this well known lyric from William Shakespeare's final play, *The Tempest*:

> Full fathom five thy father lies;
> > Of his bones are coral made;
> Those are pearls that were his eyes;
> > Nothing of him doth fade
> But doth suffer a sea change
> Into something rich and strange.
> Sea nymphs hourly ring his knell:
> > Ding-dong.
> Hark! Now I hear them—ding dong bell.

Here we can see that the poet has initially employed a ballad, or song, stanza, with the second and fourth lines shrinking to three beats and balanced by the pair of four-beat lines. The poem is rhymed in alternating lines in the first stanza (lies/eyes; made/fade), but then Shakespeare changes things up in the second stanza. He holds each line at four beats, while stretching the third line to five beats. Also, he rhymes "change" and "strange" sequentially and inserts "ding-dong" to modify the metric and rhyming patterns. This replicates the irregular boom and crash of waves, which is not as perfect in its timing as a metronome, but changes every now and then. Notice also how the poet employs assonance (vowel sounds repeated) and alliteration (consonant sounds repeated) to bring a musical quality to his lines as with the "e" sounds repeated four times in the seventh and eight lines (trees, leaves, daises, barley), and with the three "g's" and two "h's" repeated in the first and second lines (young, grass, green; house, happy). All of this is not accidental. Dylan Thomas has carefully arranged the words and the sounds the words make to create the effect of music; hence, the appellation "lyric" poetry.

Free verse has dominated poetry, especially poetry about nature, since the American poet Walt Whitman first utilized it for his long epic poem *Song of Myself.* Whitman's revolution was to replicate the sounds and cadences of everyday life and speech in poetry. Verse has never been the same since. Whitman celebrated the mundane in nature, and elevated the commonplace to the stature of the heroic, as in this famous passage from *Song of Myself:*

> I believe a leaf of grass in no less than the journey-work of
>   the stars,
> And the pismire is equally perfect, and a grain of sand, and the
>   egg of the wren.
> And the tre-toad is a chef-d'oeuvre for the highest,
> And the running blackberry would adorn the parlors of heaven,
> And the narrowest hinge in my hand put to scorn all machinery,
> And the cow crunching with depress'd head surpasses any statue,
> And a mouse is miracle enough to stagger sextillions of infidels.
>
> I find I incorporate gneiss, coal, long-threaded moss, fruits,
>   grains, esculent roots,
> And am stucco'd with quadrupeds and birds all over,
> And have distanced what is behind me for good reasons,
> But call any thing back again when I desire it.
>
> In vain the speeding or shyness,
> In vain the plutonic rocks send their old heat against my approach,
> In vain the mastodon retreats beneath its own powder'd bones,
> In vain objects stand leagues off and assume manifold shapes,
> In vain the snake slides through the creepers and logs,

In vain the elk takes to the inner passes of the woods,
In vain the razor-bill'd auk sails far north to Labrador,
I follow quickly, I ascend the nest in the fissure of
    the cliff.

Like many admirers and writers of nature, Whitman expressed an ambivalence about science and technology, as in his poem "When I Heard the Learn'd Astronomer":

When I heard the learn'd astronomer,
When the proofs, the figures, were ranged in columns
    before me,
When I was shown the charts and diagrams, to add, divide,
    and measure them,
When I heard the astronomer where he lectured with much
    applause in the lecture-room,
How soon unaccountable I became tired and sick,
Til rising and gliding out I wander'd off by myself,
In the mystical moist night-air, and from time to time,
Look'd up in perfect silence at the stars.

Equally influential on American nature poetry, particularly that of poets like Robert Frost ("The Road Not Taken"), Wallace Stevens ("The Idea of Order at Key West") and, more recently, Wendell Berry, has been the body of work left by Emily Dickinson. Unlike Whitman, Dickinson preferred to work in metered and rhymed stanzas (alternately rhymed lines of four beats, three beats, four beats, and three beats), but she sometimes improvised slightly on that form, as in poem 1755 (her poems were numbered consecutively because they were untitled):

To make a prairie it takes a clover and one bee,
One clover, and a bee,
And revery.
The revery alone will do,
If bees are few.

Like many other poets (Samuel Johnson, John Keats, Robert Frost, Robert Lowell, John Berryman, Anne Sexton, Sylvia Plath), Dickinson seemed temperamentally to alternate between a sort of manic ecstasy, as we see in 1755 with its celebration of "revery" in simple nature and a mind-numbing despair. Poem 764 is indicative of her darker mood:

> Presentiment—is that long Shadow—on the Lawn —
> Indicative that Suns go down—
> The Notice to the startled Grass
> That Darkness—is about to pass—

In her best work, the two dichotomies reach a sort of resolution, as in the very metaphysical Poem 301:

> I reason, Earth is short—
> And Anguish—absolute—
> And many hurt,
> But, what of that?
>
> I reason, we could die—
> The best Vitality
> Cannot excel Decay,
> But, what of that?
>
> I reason, that in Heaven—
> Somehow, it will be even—
> Some new Equation, given—
> But what, of that?

The repeated line "But, what of that?" becomes a musical refrain to the questions asked or assertions made in the first three lines of each stanza, and helps to bind the whole together with an ironic acceptance of mortality.

The scholar Robert Richardson has theorized that the two main lines of American nature poetry—the former from Whitman and the latter from Dickinson—arose from sources in the Colonial period. He believes that

Whitman's energetic, highly partisan free verse can be traced directly to the political pamphlets of the revolutionary period, with their improvised, often unrhymed poems and songs urging citizens to unified action against the British. Conversely, Richardson has argued that the poetry of Dickinson, with its dark undertones, its often rigid meter and rhyme, and its preoccupation with death, can be traced to the tombstone epitaphs and puritanical folklore of death in New England. Whatever the case, it is clear that most contemporary poets of nature, from John Haines in northern Alaska to Wendell Berry in northern Kentucky, can ultimately trace their verse, and their attitudes toward nature, to one of these two American icons. Any American writer contemplating poetry as a means of expressing themselves about wild nature or human nature would be well advised to begin their study with a look at Whitman and Dickinson. Given time, and practice, a poet of nature can create what the Romantic poet John Keats called "a thing of beauty," a salute to the eternal beauty of nature that becomes a thing immortal itself, as in this poem by John Haines:

"The Turning"

I

A bear loped before me
on a narrow, wooded road;
with a sound like a sudden
shifting of ashes, he turned
and plunged into his own blackness.

II

I keep a fire and tell a story:
I was born one winter
In a cave at the foot of a tree.

The wind thawing in a northern
forest opened a leafy road.

As I walked there, I heard
the tall sun burning its dead;
I turned and saw behind me
a charred companion,
my shed life.

Practice Exercises

1. Write a short imaginative piece from the point-of-view of an animal, such as a white tailed deer living in rural Virginia in a Civil War battlefield park, or a red-tailed hawk flying over the west Kansas prairie in search of prairie dog colonies, or a killer whale feeding on salmon in Alaska's Prince William Sound. In order to help get you started, here is an excerpt from a description I wrote in *Wildlife in Peril* of a river otter living in Rocky Mountain National Park, told from the animal's point-of-view.

> He swam in a slow and easy motion, with strong regular kicks of his thick webbed hind feet and even strokes of his smooth round tail, which he used as a rudder to guide himself through the water. He was cruising upstream on the North Fork of the Colorado River in Rocky Mountain National Park, pushing steadily against the current with only his eyes, nose and ears showing, beaver-like, above the surface. All his lines were sculpted and swept back, like a piece of wood that has been in the water a very long time, a once-tangled branch from which everything irrelevant has been removed, so that only the simplest of forms, the core of beauty, remains. It was a lovely slender body with a short muzzle, a broad and flattened head, a muscular neck and trunk, and a powerful tapered tail trailing behind it to a point. Everything about him was made for the water, from the head curved like a river stone, to the tiny ears flattened into the dense fur, to the fur itself—a luxurious dark brown coat so fluid and fine that it seemed

he wore the very element in which he swam. Only his stiff white whiskers protruded visibly outward from the thick muscular contour of his body, and they offered only passing resistance to the current. To the water he was just another wave gliding through the ever-flowing continuum. It was an hour before daybreak and the full August moon, still high among the spire-topped spruce and fir, brightened the winding course of the river. In the deep pools where the clear snowmelt gathered behind the rock falls or the log jams, the water was so still that the tall stars could twinkle upon it and the ghostly clouds gliding over the moon. In other places, where the channel narrowed and the water sped over the stones, the splashing water scattered the light into a myriad of steel-like slivers glimmering on the surface of the streams. The pools provided the best fishing for the otter, for in their waters the trout gathered and the suckers scavenged and the frogs jumped in now and then.

2. The tradition of poets being inspired by the prose of travel literature is a long one. William Shakespeare, for example, wrote his 1611 play *The Tempest* after reading the report of an expedition to Virginia becoming marooned on the Bermuda Island's in 1609. Several of Shakespeare's lines strongly echo the prose of William Strachey's original prose account. Strachey writes, for example, that "Once [during what he calls "The Tempest"], so huge a Sea brake upon the poope and quarter, upon us, as it covered our Shippe from stearne to stemme, like a garment or a vast cloude, it filled her brimme full . . . I thought her [the ship] alreadie in the bottome of the Sea; and I have heard [Sir George Summers] say, wading out of the floud thereof, all his ambition was but to climbe up above hatches to dye in Aperto coelo, and in the company of his old friends." Shakespeare echoed this passage in these lines from *The Tempest:* "Now would I give a thousand furlongs of sea for an acre of barren ground—long heath, brown furze, anything. the wills above be done, but I would fain die a dry death." Similarly,

William Wordworth wrote his well-known lyric poem "The Solitary Reaper" after reading a descriptive chapter in Thomas Wilkinson's 1824 book *A Tour of Scotland,* and Samuel Coleridge wrote his lyric poem "Kubla Khan" after reading, among others, William Bartram's *Travels in Florida,* and Marco Polo's *Travels in Asia.* Pick a contemporary work of travel literature, such as Bruce Chatwin's *In Patagonia* or Peter Matthiessen's *Snow Leopard,* and write a poem that features the exotic scenery, as inspired by the prose, in a way that makes the language and the place uniquely your own.

3. Compose a line or two of poetry, a stanza of poetry, or a complete poem that attempts to capture in words the sounds and rhythms of nature, such as the steady boom and hiss of the combers at the seashore, or the quiet patter of rain falling in the forest, or the uplifting music of a bird song, or the roar of a waterfall or cataract. For inspiration, listen to these lines from the poet Shakespeare, in which he attempts to evoke the complex harmony of pleasant sounds one encounters on a tropical island. Notice how skillfully Shakespeare uses the sibilant "s" to replicate the whispering sound of the wind in the leaves; a pattern of open "e" and "ou" sounds to reproduce the modulations of the waves, and a simple five beat iambic (unstressed/stressed syllable units) line to convey the musical rhythms of wild nature:

> Be not afeard. The isle is full of noises,
> Sounds and sweet airs that give delight and hurt not.
> Sometimes a thousand twangling instruments
> Will hum about mine ears, and sometimes voices
> That, if I then had waked after long sleep,
> Will make me sleep again; and then, in dreaming,
> The clouds methought would open, and show riches
> Ready to drop upon me, that when I waked
> I cried to dream again.

Chapter Twelve

Revision

I write every paragraph four times: once to get my meaning
down, once to put in everything I left out, once to take out
everything that seems unnecessary, and once to make the
whole thing sound as if I had only just thought of it.

—Adolph Murie, from the Preface
to *The Grizzlies of Mount McKinley*

Writing is a form of concentrated language, a way of knowing, a type of
perception. It is based on criteria of excellence that have been agreed upon,
in one form or another, since the guild was first established following the
invention of formalized alphabets. All writers labor, then, within this guild
and community of writers and readers to bring their works as close to perfection as possible. The process by which they edit and refine an initial
draft according to the accepted standards—universality, stylistic innovation, personal revelation, depth of feeling and craft—is known as revision.
For some writers, the process hardly exists at all. Shakespeare, for example, was known for the effortless speed at which he composed and the
almost complete absence of revision. His friend, fellow playwright, and
then-poet laureate Ben Jonson more than once paid tribute to Shakespeare

in this regard. Other writers, as was the case with Jonson himself, work laboriously to achieve the same effects. In the end it doesn't matter how the work of literature is produced—by the burning phosphorus of inspiration or on the hard anvil of revision. Posterity's concern is only for how finely the work is crafted in the end.

Generally revision begins a few days after composition, following a "cooling off" period. Thoreau expressed his thoughts on the matter at two points in his journal:

> In correcting my manuscripts, which I do with sufficient phlegm, I find that I invariably turn out much that is good along with the bad, which it is then impossible for me to distinguish—so much for keeping bad company; but after the lapse of time, having purified the main body and thus created a distinct standard for comparison, I can review the rejected sentences and easily detect those which deserve to be readmitted. (February 28, 1854)

> I find that I can criticise my composition when I stand a little distance from it—when I do not see it, for instance. I make a little chapter of contents, which enables me to recall it page by page to my mind, and judge it more impartially when my manuscript is out of the way. The distraction of surveying enables me rapidly to take new points of view. A day or two surveying is equal to a journey. (April 8, 1854)

Thoreau preferred to rest for a few days, after the completion of a draft, and undertake some work as a surveyor (one of the ways he paid his bills, such as they were). Upon returning to the text, he had a fresher perspective and could read the essay more objectively through the process of revision. What he is saying with some humor in the first quotation ("I can review the rejected sentences and easily detect those which deserve to be readmitted") is reminiscent of Samuel Johnson's famous proclamation, of which Thoreau may have been aware: "Read over your compositions and, when you meet a passage which you think is particularly fine, strike it out."

It is also probably not a coincidence that these passages on revision occurred during the period (late 1853–early 1854) when Thoreau was engaged in the most important revising of his career—the sixth and seventh drafts of *Walden.*

The process by which Thoreau revised what could be considered the world's first "nature" book is fairly well known because his journals, letters, and literary manuscripts are so well preserved. It follows a paradigm that is quite common among writers. The composition of the first 117-page draft began as early as 1846, while Thoreau was still at Walden Pond. A strategic decision was made almost immediately. "I could tell a pitiful story respecting myself," wrote Thoreau, "with a sufficient list of failures, and flow as humbly as the very gutters." Instead, he proposed "to write [not] an ode to dejection, but to brag as lustily as chanticleer in the morning, standing on his roost, if only to wake my neighbors up." This affirmation of life is at the heart of the early drafts of *Walden* and remains intact, regardless of revisions, to the completed work. The scholar Robert Richardson believes the writing and revision process for *Walden* can be divided into thirds—the early period (1846–1849), the middle period (1852), and the last period of drafts five, six, and seven (1853–1854). Walden Five was, according to Richardson, the "major reshaping" that turned the book into something like what we now see, with the narrative divided "into separate chapters with chapter headings" and the section on autumn given "its proper weight." Revisions in Walden Six and Walden Seven consisted primarily of additions, such as the important clay-bank passage in "Spring" and the humanizing anecdotes about his neighbors—the freed slaves and poor Irish squatters—in various chapters. Above all, Thoreau was concerned with the maintaining the unity of the overall structure of the book:

> *Walden* follows the seasons of the year, and Thoreau's revisions show his careful attention to this structural center, but he was at the same time working to emphasize other unifying aspects of the book, such as the concept of the imagination, the sense of the sacred, the idea of a prose poem of work and the earth, the double concept of nature as both scenery and force, the transcendental center represented by the chapter

"Higher Laws," and the practical, ethical imperative of "Conclusion."

It is hard to imagine today, but important to remember in considering the revision process that Thoreau was attempting something that had never been done before. Naturalist's calendars had been written since the days of the Romans (Varro, Columella, Cato), but no one had yet written a philosophical treatise in the form of a series of seasonal essays. Richardson further notes that Thoreau toiled over the page proofs for *Walden* all spring and "was an interminable reviser [who] could not stop improving his pages even after they were out of his hands." He quotes Thoreau on the latter's passion for last minute revisions:

> When I have sent off my manuscripts to the printer, certain objectionable sentences and expressions are sure to obtrude themselves on my attention with force, though I had not consciously suspected them before.

The important thing with revisions in the page proof stage, as we shall see in the last chapter of this book, is to keep those alterations to a minimum, as author's are charged for every change to the text beyond a certain limit.

Generally speaking, revision is concerned with two areas: adjustments at the level of the sentence and paragraph (micro-level) and improvements in the overall structure (macro-level). There is no set rule, of course, nor should there be, for which to undertake first. Revision is of necessity a somewhat haphazard process, with a great deal of improvisation and adaptation involved. Picture a sculptor slowly working with a piece of stone in his or her studio, chipping away at the nonessential parts and smoothing over the rough edges to reveal the hidden, perfect figure within. It is safe to say that during this period writers look at their work from virtually every angle possible. They also read it from the perspective of others—the editor, the publisher's sales representative, the bookstore owner, the audience of today, the audience of tomorrow, the reviewer, the teacher, the friend, the colleague, the reader half way around the world who knows nothing about the subject (perhaps the most important imaginary reader if you are to universalize your particular experience)—in an attempt to obtain that critical distance, that objectivity, that will enable them to approach a text they have created as freshly as possible. I sometimes read over an essay or chapter literally dozens of times, each time sifting through the text anew to make certain each word was perfect for the position it occupied in the sentence. During this period the author considers many things: consistency (especially of diction, style, imagery, symbolism, and motifs), concision (the fewest words possible to express ideas), mechanics (punctuation, spelling), grammar (misplaced modifiers, subject-verb agreement, parallelism, and so forth), sentence rhythm and variety, opening and closings, transitions (between sentences, paragraphs, and paragraph blocks), strength of argument (in the case one is being made), paragraph coherence, and structural balance and unity.

Before you send the essay or book off, it might be a good idea to put it aside for a few days, perhaps a week, if you can. When you return to the work, you might ask yourself some questions:

1. Is the opening as effective as I can make it?
2. Is the closing as effective as I can make it?
3. If I am making an argument, have I made my case convincingly and refuted all arguments to the contrary?
4. Are my transitions smooth?
5. Are there any points that require additional development?
6. Are there any places where I might make further cuts?
7. Will the reader understand the intended meaning?
8. Have I explained where all quotations come from, either through attribution in context or through citation?
9. Is this absolutely the best I can do?
10. How will this read to a person living a century from now?

Hemingway once said with regard to revision, and with the bluntness he was known for, that "the most essential gift for a good writer is a built-in shock-proof, shit detector. This is the writer's radar and all great writers have had it." In a sense, that is what revision consists of—removing with the delicacy of optical lasers, or, if necessary, with the thoroughness of McCulloch chainsaws—anything that obscures meaning, uselessly adorns, or unnecessarily calls attention to itself. What remains is only that which serves a purpose, which justifies inclusion, like the antler's on a deer, or the feathers on the eagle, or the tooth on the wolf. You are creating something that will, if the job is well done, have life, and proper revision insures it will have a long and productive life, and not be torn apart by sharks or drown in the first storm.

There is an old story that the Roman poet Virgil, on his death bed and seeing that his twelve book epic poem *The Aeneid* was still short a few hundred lines, ordered that the manuscript be burned. Somehow or another this request was communicated to his patron, Octavian Augustus, who had commissioned the poem to be written as a celebration of Roman culture. Octavian, of course, dispatched his centurions to seize the work, and after reading it was glad he did. The emperor realized that Virgil had crafted a literary creation that would insure that Rome would live on even when its great families, powerful armies, vast cities, famous lawyers, and fabled daughters were dust. The point is that artists are rarely fully satisfied with their work. They have a standard of excellence that supercedes any that a mere guild could impose. They are writing not for the morning paper or the periodical magazine, but for the ages, for mornings that will not arrive for two thousand years, for men and women who will inhabit a world whose science we cannot even imagine, but who will value excellence, truth and beauty as we do.

Raymond Carver, who once said that any of his good short stories took at least "twenty or thirty drafts," was fond of telling the story of how Leo Tolstoy revised *War and Peace* even when the work was in galleys and pages:

> Its instructive, and heartening both, to look at the early drafts of great writers. I'm thinking of the photographs of galleys belonging to Tolstoy, to name one writer who loved to revise. I mean, I don't know if he loved it or not, but he did a great deal of it. He was always revising, right down to

the time of page proofs. He went through and rewrote *War and Peace* eight times and was still making corrections in the galleys.

Tolstoy probably did not enjoy revising anymore than he did going to the dentist, but he felt obligated to revise for two reasons—to fulfill the contract he had with his audience to produce a work of excellence, and to honor the covenant he had with his work to make it as good as he possibly could. These are solemn obligations, not entered into lightly and not discharged easily. It is probably not an overstatement to say that one of the most important parts of any literary work is that which has been removed. The author has eliminated, often with great pain at parting with something loved, what was extraneous. We are left only with what is essential. So, too, does the process of natural selection constantly work on organisms, adjusting and improving, refining and revising. The process of revision is, then, at one with the deepest processes of nature, and so a nature writer, laboring late at night over a particularly stubborn closing paragraph, should take some consolation in the fact that he is partaking in an activity as ancient as life on the planet. It is the force, revision, that drives evolution, that makes possible the hand that writes the words. The words compete, even as the genes, for the chance to endure. Most will not, but a few will adhere to the structures of meaning by which we preserve language. These things we call books, which are as intrinsic to culture as DNA is to the cell.

Practice Exercises

1. Keep all of your drafts for an essay. Assuming they are of a manageable number, arrange them on the living room floor. Study them closely. What sort of patterns do you see emerging? How do the openings and closings change through successive revisions? Does the overall structure remain intact? Does the climax move in position as the essay evolves through the revision process? How does language (especially syntax and diction) change as the essay is pushed toward completion?

2. Call or write a publishing nature writer and ask if he or she will save the drafts of one of their published works. Study these drafts as you did your own in exercise one.

3. If you live near a university library, contact the archival staff and ask if they have the papers of any prose writers (frequently, authors donate their papers to libraries toward the end of their lives because of the tax benefits that accrue from the transfer). Assuming that drafts of various essays or books are in the collection (as with Edward Abbey's papers at the University of Arizona, Tucson or Adolph Murie's papers at the University of Alaska, Fairbanks), ask permission to study those drafts with an aim toward understanding more fully how this successful writer worked with language through the process of revision.

Chapter Thirteen

Research

. . . as a writer I am dependent on scientific inquiry for information. If I am going to write coherently—about polar bears, for
example—I am dependent upon the scientists who work with polar
bears for solid information of a certain sort. And yet I am troubled
by this because of the ways we approach animals as scientists.

—Barry Lopez,
from a discussion with Edward O. Wilson on "Ecology and the Human
Imagination" held at the University of Utah on February 1, 1988

My introduction to research came with a book entitled *Wildlife in Peril: The
Endangered Mammals of Colorado*. In its original conception, the book was
intended to focus only on the grizzly bear and the gray wolf, the former a
possible relict population and the latter an extirpated species in Colorado.
My editor and publisher, however, insisted that I expand the proposal to
include all six federally listed endangered mammals in the state. He believed,
rightly, that this would make the book more worthwhile for readers and
also more beneficial for the cause of endangered species in the southern
Rocky Mountains. The expansion would, not incidentally, generate more
reviewer interest and increased sales. The widened table of contents brought
the river otter, the wolverine, the lynx, and the black-footed ferret into the

fold. While I was generally familiar with the contemporary scientific literature on the wolf and grizzly, I knew essentially nothing about the other four.
What ensued—nine months of intensive reading, interviewing, traveling,
drafting, editing, and revising—presents a classic case study of how a nature
writer researches a largely new subject, exploring sources, studying data,
analyzing contrary views, translating specialized discourse, and eventually
synthesizing the whole into a work of his own.

The first stage of this project, as with any book project, was to make
a detailed list of the jobs that needed to be done and carefully determine
what the production schedule would be. In the business and government
world they call this a time management plan. The production schedule here
consisted of a complete outline of the book—its chapters, or essays, and
their constituent parts. Each chapter had four parts to it (introductory
sketch, past history, natural history, current status and future prospects for
recovery), and the six mammal chapters were parallel, or identical, in construction. Three additional chapters covered such topics as the fossil history of mammals, the arguments for preserving endangered species, and a
sort of manifesto of what needed to be done statewide in a practical sense
to save the six animals (and those at risk for becoming endangered). After
some analysis, I determined that the 220-page book would require around
720 man/hours of work, or about twenty hours a week for thirty-six weeks
in order to complete it by the deadline stipulated on the contract (it takes
me over-all about three hours to write one page—one hour for research,
one hour for writing, one hour for revision). I was working a full-time job
at the time, and so I had to carefully chart out each week to allow for those
twenty free hours. Once I had divided the rather intimidating 720 hours
into smaller jobs of four or five hours, the book became simply a matter of
choosing from an interesting array of assignments each time I sat down to
work on it. Before I knew it, the first draft was completed and the manuscript was ready to be scrutinized by the numerous technical reviewers the
editor had arranged to review the work for accuracy. I have applied this
same system to every book project (sixteen books) since and have found
that it always works perfectly.

Having completed the production schedule, the research phase of the
book project then began in earnest. The research phase was divided into

two broad categories: city work and fieldwork. The fieldwork was the most fun, of course, and involved traveling to different parts of the state, talking with local people and wildlife people, and hiking in and photographing habitat. Let me give one example to illustrate the process. For the grizzly bear chapter, I drove eight hours south of Denver to the south San Juan Mountains. While there, I hiked twelve miles up the South Fork of the Conejos River to Blue Lake, where a grizzly bear had been killed on September 24, 1979; interviewed Ernie Wilkinson, who had killed a grizzly bear on the upper Rio Grande River in 1952; interviewed Art Dimeo, assistant district ranger for the Rio Grande National Forest; stopped by the Platoro Lodge to view the bear killed by Wilkinson; interviewed sheepherders and wranglers to see if they had observed grizzly bears recently (they had); visited the Banded Peak Ranch on whose property the last grizzlies were presumably living; and spoke with landowners (cattlemen and woolgrowers) in the area to solicit their views on the controversial species. This same process was repeated in different parts of the state—the Trappers Lake region, the North Fork of the Colorado River, the Rio Blanco country, the Pawnee National Grasslands, the Vail Pass area—for each of the six endangered mammals.

The city work was less exciting than the fieldwork, but was ultimately more important to the success of the project. Here I acquired the facts and data necessary for me to accurately present my plan and convincingly make my arguments. Let me again use the grizzly bear as an example. The first step was to visit the non-game office of the Colorado Division of Wildlife in Denver. I interviewed the non-game manager about the species and then sat down with the approximately 4 linear feet of material in the grizzly file and looked at every page (skimming whenever common sense dictated). She was kind enough to permit me the use of her copy machine in order to reproduce essential historical documents and reports in the archive. The next step was to visit the Federal Center and spend some time at the U.S. Fish and Wildlife Service Library, which, again, contained valuable historical information on the species in the state and region; while at the Federal Center I also interviewed individuals in the Public Affairs office and Endangered Species Office for the Agency. During my interview with the head of the Endangered Species Office, I learned that the Denver Natural

History Museum had the skull and pelt of the grizzly at Blue Lake in its permanent collection, and so I then visited the museum and spent some time studying the various skulls and pelts and interviewing the·curator of mammals. She recommended that I visit the zoo just down the street and observe the captive grizzlies there and talk with a professor in Boulder who had worked with grizzlies in Yellowstone, and I later did both. The next phase of research for this chapter involved interviewing key players in any potential attempt to reintroduce the bear: the Director of the Colorado Division of Wildlife, the Executive Vice President of the Colorado Cattleman's Association, the Public Affairs Coordinator for the Colorado Woolgrower's Association. I also spent some time interviewing (by phone) prominent bear biologists such as Chris Servheen, the recovery team coordinator for the species for the U.S.F.W.S., Dave Brown, a biologist with the Arizona Game and Fish Department, Al LeCount of the same agency, and John Craighead in Missoula. The final part of the research phase for this chapter consisted of dozens of hours spent immersed with published articles, monographs, and books, ranging from Chris Servheen's dissertation on grizzly bear denning ecology in Montana's Mission Mountains to Marcel Couturier's classic study of the French brown bear *L'ours Brun* (in which all those years of studying French in grammar school, high school and college were finally put to good use). This process was repeated, in essentially the same form, for each of the six animals in the book.

One of the most difficult aspects of the project for me was when two scientists with equally respectable credentials on a particular species disagreed, sometimes strongly, on a specific issue. For example, Dr. Chris Servheen, the federal biologist who oversees all grizzlies in the Lower 48, believed that the genetic effects of interbreeding would doom the "island population" of grizzlies in the south San Juans. He later published a scientific paper in which he argued for the necessity of periodically augmenting isolated populations to maintain genetic viability. Dr. Tom Beck, on the other hand, the state biologist who led the 1980–1982 search for grizzlies following the 1979 incident, believed that the genetic effects of interbreeding would be minimal, and pointed to the bear populations of France, Spain, and Italy as proof that bears could interbreed for hundreds of years and not self-destruct genetically. Similarly, just about every biologist I interviewed had a different opinion on whether the bears remained in Colorado, whether the bears could be restored, where the restoration would occur, and in some cases there was disagreement as to fundamental aspects of grizzly bear ecology, such as predation, diet, denning behavior, and even what constituted definitive sign. What I finally decided is that, in most cases, my readers expect and trust me to make an informed judgment on the issue, with the realization that there are often more advantages than disadvantages to being an outsider in a contentious situation. On several occasions, however, I presented both sides of the argument so that the reader could make up his or her mind on the issue. For example, I quoted both Tom Beck and John Craighead on the fundamental issue of whether grizzlies should be restored in Colorado. They disagreed totally on this issue (Beck against and Craighead for), and I wanted the reader to understand how divisive the issue was, even among respected experts in the field who one might expect would have some fundamental agreement.

A few basics on the mechanics of research. First, always use a tape recorder while interviewing. Most retail electronics stores sell a device for $25 that enables you to record phone interviews (let the person on the other end know they are being taped solely to insure accuracy). As a courtesy, you should always send manuscript pages or galley pages to those interviewed for proofreading. While this usually is not possible for journalists, who are operating under a deadline, the long production schedule for a book enables you to avoid embarrassment or hurt feelings upon publication. Second, always use note cards and bibliographic cards when copying from sources; the former should include quotation marks to prevent confusion as to sources and the latter makes citations (if necessary) or a bibliography much easier. Third, avoid general or soft sources like encyclopedias, popularizations, and unreliable informants as they weaken your credibility as a professional writer with reviewers and readers. Fourth, learn the value of librarians in your research; they are amazingly well-informed not only as to their own collections but also as to other resources you might be unaware of. Similarly, learn how to use inter-library loan to acquire rare or hard-to-find articles or books from collections around the country or world. Fifth, become effective at using brackets and ellipses to smoothly integrated quotations into text; these two devices enable you to alter verb tense and grammatical construction in order to incorporate a sentence gracefully into its new context. For example, here are two block quotes from *Wildlife in Peril* in which I used ellipses and brackets to make a source fit more smoothly in the text:

> The remoteness of the area [south San Juans], its proximity to wilderness areas, and the existence of a very large . . . Spanish Land Grant [the Banded Peak Ranch] all lend credibility to the possible existence of a relict population.

> [There is] no question that the physical habitat for grizzly bears exists in Colorado. That really isn't an issue. What is an issue is 1986 mankind's intolerance for grizzly bears in Colorado. Change the attitudes first before sacrificing the bears [in a recovery effort]. My feelings are I have never known of a grizzly bear I disliked enough to turn loose in Colorado.

In the first case, I used brackets to explain references in the text more fully to the reader, and ellipses to eliminate some unnecessary words in the sentence. In the second case, I used brackets to turn a sentence fragment into a sentence with a subject and a verb and to complete a sentence whose meaning suffered when excised from its source.

One of the most important things is to have a working research library at home, and/or to have access to research materials through the Internet. You simply cannot be running off to the library every time you need a fact. A home library is not something you can assemble overnight, unless you are very wealthy and most writers are not. My basic research shelves consist of around one hundred books, which represents about 3 percent of my total collection, took me years to put together. These are the books I have within easy reach of my computer table, and I am using them constantly. The most important titles include the following:

1. *The Oxford English Dictionary* in two volumes
2. A thesaurus

3. Field guides to all major animal and plant groups for the U.S., Africa, and Asia

4. Feldhammer and Chapman's *Mammals of North America*

5. An encyclopedia of world art

6. An encyclopedia of film

7. A world atlas

8. A North American atlas

9. A dictionary of mythology

10. A dictionary of folklore

11. An atlas of archaeology

12. A current almanac

13. Current textbooks for the basic sciences

14. Clarence Glacken's *Traces on a Rhodian Shore* and Roderick Nash's *Wilderness and the American Mind*

15. A biographical dictionary

16. The Oxford dictionary of quotations

17. The *Norton Anthology of English Literature* in two volumes

18. The *Heath Anthology of American Literature* in two volumes

19. An anthology of world literature in two volumes

20. *Chicago Manual of Style*

21. *The Timetable of History*

22. The complete works of several literary figures (Shakespeare, Milton, Thoreau, Solzhenitsyn, Mishima, Kawabata, Camus, etc.)

23. *A Literary History of the American West,* edited by Tom Lyon

24. Grammars and dictionaries for the three languages I can translate fairly well—French, German, and Anglo-Saxon—as well as grammars and dictionaries for Latin, Greek, Chinese, and Japanese

25. Major religious texts

26. Current USFWS recovery plans for endangered mammals

27. A textbook on surgical anatomy

28. The Durant history of civilization
29. A textbook of American history
30. Various works of world mythology

This may seem an eclectic collection of primary research texts, but every one of these books has been used at least once in the past month, and so they stay on my most active shelves, always within easy reach. There is nothing more frustrating than to be in the middle of some complicated project, facing a deadline, and not be able to find a key quotation or fact. One day soon we will have everything available on our computers, but until that time, I would advise building the best home research library you can.

Research is equally important to novelists who adopt nature as a theme. While writing his historical novel *The Big Sky* in 1946, for example, A. B. Guthrie spent a year at Harvard as a Niemann Fellow. At Widener Library he read whatever he could find on the Missouri River of the 1820s and 1830s and more or less constantly conferred with frontier historian Bernard DeVoto. Several obscure novels written during the early nineteenth century by Scottish adventurers and hunters provided Guthrie with the unique dialects that were spoken by his chief characters Boone Caudhill, Jim Deakins and Dick Summers. Although Guthrie was originally from western Montana, and was familiar in a general sort of way with the history and geography of the fur trade, it is doubtful he could have written such a successful novel without having conducted the intensive research at Widener Library. Similarly, John Steinbeck wrote his novel *The Grapes of Wrath*, a novel in which nature is very much the central character, only after having spent six months in the California camps where the displaced Oklahoma farmers lived. During that time he interviewed the Dust Bowl refugees, wrote a number of lengthy newspaper articles, and actually came to know the people who the world now knows as the Joads. It is possible, of course, to write a work of fiction completely from the imagination, but those novelists who conduct research in a thorough and professional way probably have a better chance of creating a comprehensive reality in their fiction, subsequently attracting readers and reviewers, and ultimately being recognized with literary awards like the Pulitzer and Nobel (as with Guthrie and Steinbeck).

If the research that went into these books, or into *Wildlife in Peril,* sounds like a lot of work, it is. But it was also fun in my case, because I was learning about animals I loved, and because I realized that my little book, if researched thoroughly and written well, could improve the state of affairs for these beleaguered species. There is a genuine pleasure in speaking to people who are at the center of research activity on a particular subject, in suddenly spotting the data in a lost study that will prove a point that could change the outcome for a fragile population of animals, in walking to a wild windy place high in the mountains that you have seen described only in dry summary accounts and incident reports, in touching the bleached skull and tanned pelt of the legendary animal you have read about only in books, and feeling the awful sadness of extinction. Many of the contemporary nature writers whom I most admire are those, like George Schaller, Edward O. Wilson, Barry Lopez, Richard Nelson, Jane Goodall, and Cynthia Moss, for whom research, both in the library and the field, is a passion. Few tasks are more demanding—the tedium, the distractions, the sheer workload—but for those who approach the challenge with vigor and good humor, professionally executed research can result in books that not only are read, but that endure.

Practice Exercises

1. Pick an interesting or unusual animal that lives in your home area (e.g., the armadillo in Texas, the copperhead in Ohio, the wood frog in Alaska, the San Joaquin kit fox in California). Visit your regional state wildlife office or the nearest university or college library and gather whatever materials are available on the species. This should include such sources as scientific journal articles, government field studies, and published books. Immerse yourself in these resources for several weeks, prepare an outline, find an opening into the topic, and compose a natural history essay that translates the dry specialized discourse and presents the animal in a fresh new light. Translating specialized discourse can be a daunting task. Here is an excerpt from the Grizzly Bear Recovery Plan, prepared by the U.S. Fish and Wildlife Service:

> Servheen (1980), using the most conservative estimate for
> the NCDGBP population (1/30 miles exclusive of Glacier
> National Park), computed an initial estimate of adult
> females (4.5 years and older) in this ecosystem. From a
> sample of 180 bears of known sex and age (kills),
> Servheen (1980) develop a survivorship curve which indi-
> cated 19% of the population were adult females . . .

For the second part of the exercise, send this essay to at least one expert on
the species, have the essay thoroughly critiqued, and then revise the essay
accordingly.

2. Repeat the procedure previously described, but instead of an animal
    choose a plant, such as the Saguaro cactus in Arizona, the Ponderosa
    pine in Montana, the Live Oak in Louisiana, or the orange tree in
    Florida. For ideas on how to approach this topic, read Henry David
    Thoreau's lovely essay "Wild Apples," which is not as well known as
    some of his other essays, but which is nevertheless one of the best he
    ever wrote.

3. Visit with a wildlife researcher at a government agency or university
    with the object of interviewing this individual, familiarizing yourself
    with her or his research, and writing an essay on their work. If you
    lived in Colorado, for example, you might find yourself writing on
    Tom Beck, who is currently undertaking a long-term study on black
    bears for the state in order to determine if overhunting is taking place;
    in Minnesota you might consider any one of a number of wolf
    researchers employed by the U.S. Fish and Wildlife Service and
    National Park Service who are studying predator/prey dynamics and
    the effects of wolves on livestock; in Florida your time might be well
    spent with state biologists studying the endangered panther. (Note: this
    sort of writing makes excellent features for local newspapers, particu-
    larly in Sunday magazine sections; it would be a good idea to take a
    camera and obtain as many images as possible). Brainstorm with your
    friends and colleagues as to the best questions to ask.

4. Locate two scientists who disagree on an important research or management issue involving wildlife. Familiarize yourself with the relevant facts and studies, interview the two opposing experts, and then write an essay in which you sort out the arguments and take a stand with respect to the central questions. Remember, that your loyalty is not to a particular person, government agency, or nongovernmental organization, but rather to the truth, which is what your readers count on you to provide them.

5. Research a particular river, forest, mountain, or landform in your home region and write an essay that places it in an historical as well as a biological context. For example, if you lived in Denver you could write an essay on Longs Peak, located in Rocky Mountain National Park and named for Major Long, who conducted an 1819–1820 expedition to the headwaters of the Platte and Arkansas River. In this essay you would want to discuss in particular the efforts of Enos Mills, who built a lodge at the base of the mountain, to protect the region as a national park; the sixty-plus climbers who have died on the mountain; and the current management problems facing the national park as visitor use burgeons. It would be useful to visit the national park headquarters in Estes Park in order to review archival material and interview the park historian. A hike to the summit would also be a good idea, in order to talk to people on the trail and gather the sorts of personal anecdotes that make an essay come to life. In fact, such a trip narrative could form the structure for the essay, with the historical information intercut in appropriate places.

6. Visit a place that you have read about in a monograph or a nature essay. Compare what you observe with that the representations of it you previously encountered. Were the descriptions accurate? Did they act as a lens to clarify landscape, people, or issues, or not? Try to write your own essay on the area, improving on what was done earlier. If you live in the east, you might wish to visit the Smokies, Okefenokee Swamp, or the Everglades, reading material you pick up in the Visitor's Centers of the respective parks and comparing that with what you

encounter on your hikes. In the Everglades, for example, you could read the U.S.F.W.S. recovery plan for the Florida panther, and compare what you read there with what you find in your hikes and canoe trips into the backcountry of the park. Similarly, in the far west, any of the large national parks would provide an excellent location for this exercise. In Arches National Park, for example, you could read Edward Abbey's *Desert Solitaire*, in Yellowstone you could read the U.S.F.W.S. grizzly bear recovery plan, and in Denali you could read Adolph Murie's book *A Naturalist in Alaska*.

7. Study a particular work that effectively incorporates quotes into the text. Pick a similar subject, gather primary source materials, and follow the same techniques used successfully in your paradigm work. Good works to reference include Barry Lopez's *Arctic Dreams* or *Of Wolves and Men* and Thomas McNamee's *The Grizzly Bear*.

Chapter Fourteen

Workshopping

Writing instruction is something that did not exist in our
colleges until the 20th Century. So far as I know, the other
countries where it occurs have copied it from us. In some
countries it still doesn't exist . . . By and large, a good
writing class functions like a form of publication. Abruptly,
this manuscript—this thing that was a scribbled page—is
put into a posture of dignity, demanding attention.

— Wallace Stegner,
On the Teaching of Creative Writing

I took my first creative writing workshop in 1970. I was sixteen and studied creative writing for six weeks at Mount Hermon, a private secondary
school in north-central Massachusetts. My teacher was Paul Smyth, a poet
and recent graduate of Harvard who was teaching full time at Mount
Holyoke just down the river. He taught the class in order to help pay off
his new Harley-Davidson motorcycle and to put formula in the cabinet for
his new baby. We studied both prose and poetry that summer and he introduced us to writings as varied as Ernest Hemingway's "Big-Two Hearted
River" and Wallace Stevens "Thirteen Ways of Looking at a Blackbird."

The class was small, about a dozen, and consisted mostly of the teenage sons and daughters of affluent parents from places like Bucks County and Westchester County. I was shy and strictly middle class and from the Midwest and didn't say much in class. I did a lot of listening and watching. One day in conference Paul Smyth took a line that I had written comparing the full moon to a cream poppy with the word "like" and changed the line to read "the full moon rose, a cream poppy, over the fields." Metaphors are always more powerful than similes, he explained. Another thing I remember from that summer is his insistence that we commit three poems to memory every week (poems I can still recite); Smyth believed that only through the study of poetry does the writer learn to appreciate the value of a single word. Finally, he provided encouragement. The importance of positive reinforcement cannot be overemphasized, especially in the early stages. It would be another fourteen years before I published my first book, but Paul Smyth had planted the seed, and it put down some long roots that summer on the green forested hills above the Connecticut River.

Later I had the good fortune to study under two more capable workshop teachers—the Shakespearean scholar and poet Charles Squier at the University of Colorado and the novelist and short story writer William Wiser at the University of Denver. It took some talking on my part, but Squier finally consented to let me enter his "instructor permission required" upper division workshop during my first semester at Boulder. Every week we, and there were nine of us, wrote something new and presented it to the class. Most of the classes, at least until the snow began to fall, were held outside Hellems Hall in the Mary Rippon Shakespearean Theater. By the second week in December we each had a portfolio, which Squier insisted we try to get published. None of my work was accepted, of course, but I was introduced to the concept of publication. Fifteen years later (like most writers I wanted some courses in the university of life before resuming my formal education), I enrolled in William Wiser's graduate fiction workshop in Denver. Wiser, who had never attended college but who had published so widely they tenured him on that basis, provided me with advanced guidance on such points as outlines, narrative technique, and revision. The seven members of his beloved workshop have all, with one exception, gone on to publish books. I wrote a wretched novel that year about a rogue grizzly bear and the Ahab-like soul he wounded and Wiser dutifully and copiously red-marked the chapters. I suppose most everyone has to experience at least one failed novel in order to understand the genre (Faulkner's *Mosquitoes*, Hemingway's *Torrents of Spring*, A. B. Guthrie's *Murder at Moon Dance*), but it certainly helps to have someone working with you who has published books and is familiar enough with the terrain to be a helpful guide.

From 1988 through 1994 I taught both the nonfiction prose workshop and the graduate professional writing workshop at the University of Alaska. The graduate workshop was small and focused primarily on thesis work. Because the undergraduate course was popular and generally attracted over twenty students, I divided the class into two smaller groups, with all twenty or so meeting for a discussion of readings on Monday and then the two smaller groups of around ten each gathering together separately on Wednesday and Friday. Early in the course I identified one specific goal for each student, with the idea being to solve at least that one narrative problem over the course of the semester. The course was divided into five thematic modules, with the last three-week module (my favorite of course) devoted to nature writing. By that time—mid-April—the snow was melting rapidly and the birch trees were beginning to bud and so everyone was in the mood to write about nature. Students were free to choose their own topics within the modules. In the rare case where the student could not generate a topic, I suggested topics. The Wednesday and Friday meetings tended to be lively and fun affairs, as the class members presented their work for critique, and received comments and constructive suggestions from their peers. In the undergraduate workshops, I provided as much positive reinforcement as I could where it was genuinely deserved; in the graduate workshops I also offered positive reinforcement, but by that point students have projects further along and are generally more prepared to benefit from close critical attention.

Wallace Stegner, who for decades oversaw the writing program at Stanford University, published a little book years ago entitled *On the Teaching of Creative Writing*. Anyone contemplating a workshop degree, or teaching in a workshop program, should read this book. Stegner taught such writers as Larry McMurtry, Thomas McGuane, and Edward Abbey in his program, and his is a voiced to be listened to. Stegner places the birth of the workshop idea in the 1930s at Harvard with Dean Le Baron Russell Briggs who "began teaching a [composition] class that required a daily theme." (Stegner may have been unaware of the fact that as early as the 1830s Professor Channing required such papers from his Harvard students, one of whom was Henry David Thoreau). Later, Charles Townsend Copeland, a colleague of Briggs's, followed suit and by the 1940s the university had instituted what has since become known as the "Briggs-Copeland Faculty Instructors of English Composition." At around the same time Iowa State University began offering a Masters of Arts degree in creative writing; in fact, Wallace Stegner was one of the first recipients of this degree and eventually returned to teach at Iowa for awhile (see his novel *Crossing to Safety* for a fictionalized treatment of his tenure there). After World War II, as tens of thousands of GIs returned to school, creative writing programs began appearing faster than fireweed in a burned forest. One of the first in this period was Allan Swallow's program (1947) at the University of Denver, and soon others appeared at schools as diverse as the University of Pittsburgh and Knox College. By 1970 there were more than fifty, including the stellar programs at the University of California, Irvine and the University of Montana, Missoula, and since then the growth of these programs has been astonishing. At this writing (2003) there are over three hundred writing programs in the United States and Canada. One of the most exciting recent developments has been an organized movement to institute creative writing programs at the historically black colleges and universities—nature writing could definitely benefit from this development.

The workshop method has met with some criticism over the years, particularly from literary academics (in some departments one can observe a sort of "schism" between the literature and the writing faculty). Scholars, who spend their careers studying literature in the past tense, as opposed to works in progress, have faulted the method primarily for applying the "assembly-line" paradigm so prevalent in American culture and education to the creative arts. Their chief concern is that the result will be a homogeneity and mediocrity of product. They have also witnessed an enormous shift in graduate student enrollment from literature to writing degree programs. Even among creative artists there has been some dissatisfaction with the current approach. A. B. Guthrie, for example, who won the Pulitzer Prize for his novel *The Way West* and an Oscar nomination for his screenplay to the film *Shane*, had this to say:

> With notable exceptions college courses in creative writing are of little avail. Too often the teachers themselves are frustrated writers and at odds with the craft. Too often . . . they leave the student as lost as before. And sometimes, out of their own failures, they are sour at the prospect of student success. Where to go then? I vote for good writer's conferences, of which there are several [Guthrie participated as a student at the Bread Loaf Conference in 1946 and found a publisher there for his first serious novel *The Big Sky*; he later taught there with friends like Wallace Stegner and Robert Frost].

The writer's conference can be a useful alternative, or additive, to the university workshop for two reasons: the conference can bring the novice writer into close contact with successful professional writers, and the conference can also sometimes facilitate meetings with editors. The former can result in valuable guidance that will improve writing and the latter can benefit publication. The disadvantage of conferences is that they can be very expensive, especially for struggling writers who are not often among the wealthiest members of society.

The other alternative, and one that can really work well, is to form your own conference or workshop. This is what the writers of eighteenth-century England did—James Boswell, Samuel Johnson and others—when they gathered in the coffee houses of London to discuss literature and share manuscripts. Later, we see Ernest Hemingway following the same model with Gertrude Stein and F. Scott Fitzgerald in the bars, restaurants, and salons of Paris. There is nothing new about striking out on your own, and creating your own "school" of writers, and this is probably more popular than is generally acknowledged. A number of years ago, for example, a handful of writers from my graduate professional writing workshop saw so much progress in their writing that semester they did not want it to end. So, they formed their own private workshop group, and began meeting twice a month. Three years later, these writers still gather regularly, critiquing each other's work, discussing books they have exchanged, encouraging each other when needed. About two decades ago, I had a kindred experience while writing my book *Wildlife in Peril*. Because the book was to be sold at visitor's centers in various national parks, which meant it had to pass a rigorous review process, the publisher suggested the text be examined by a number of professional scientists. The scientists formed a sort of ad hoc workshop group as they provided feedback on an early draft of the manuscript. I must say that the initial reading of their comments was one of the most depressing experiences of my life. Scientists do not sugarcoat their responses as workshop members do. They bluntly circle and frankly annotate in red every inaccurate fact, poorly written passage, or illogical argument. Because that book meant so much to me, I swallowed my pride and began the long hard task of rewriting. I can honestly say that their constructive criticism improved that work, and my writing in general, one hundred percent. I cannot recommend such an experience too highly.

It is important to remember, as Wallace Stegner observed in the chapter epigraph, that organized creative writing workshops, as a part of higher education, did not exist prior to around 1930. Such courses and programs of study are essentially nonexistent today outside the United States and Canada. Writers elsewhere still become members of the guild the old-fashioned way—through journalism (as did Mark Twain, Ernest Hemingway, John Steinbeck, A. B. Guthrie, Edwin Way Teale), college teaching (as did Norman MacLean, Wallace Stegner, Aldo Leopold, Sigurd Olson, Joseph Wood Krutch) or some other line of work (Rachel Carson and Archie Carr's careers in biology, Ed Abbey and Doug Peacock's employment as park rangers and fire lookouts, Gustav Eckstein and Lewis Thomas's training as physicians). Literature flourished for thousands of years without creative writing programs, and if workshops ended tomorrow in the United States and Canada, literature—poetry, fiction, nonfiction—would still be created in huge quantities. Books would still be bought and sold, reviewed, talked about, placed in libraries, adapted into films, taught at universities, remaindered out as unsaleable or made into classics by popular acclaim. Novice or apprentice writers should be reassured, in an age and country in which they may be made to feel outsiders if not part of a university program, that a degree in creative writing and a comfortable sinecure do not automatically bestow legitimacy or excellence upon one's writing. In fact, many actively publishing nature writers—Rick Bass, Jim Harrison, David Rains Wallace, Barry Lopez, Jan DeBlieu, Gretel Ehrlich, Terry Tempest Williams, Linda Hasselstrom, Brenda Petersen—have no affiliation with universities, and only occasionally attend conferences.

In the final analysis the decision as to whether or not to study for a B.A. or graduate degree in creative writing is a very personal one. For some nature writers it may be quite helpful and provide a sense of community and greater self-confidence. In my case it was most beneficial in that it brought me into close contact with such stimulating thinkers as Robert D. Richardson, the biographer of Thoreau, and William Wiser, the novelist and short story writer. For others, though, it could have a negative effect on their talent and career. I can not imagine, for example, confining the restless spirit of John Muir in his twenties or thirties to a classroom, and of course one of the most notable critics of organized education was Henry David Thoreau, who once joked to Ralph Waldo Emerson that Harvard taught "all the branches" of learning but "none of the roots." Thoreau, in fact, learned writing primarily by studying the literatures of other languages; he was fluent in seven languages, including Latin and Greek. We might say that Thoreau was a multiculturalist a century-and-a-half before the term was invented. Whatever choice you make as to training—to enroll in a workshop program, to form your own workshop, to attend a conference, to study on your own—the success of the endeavor will in large measure be determined by your possessing a good work ethic and a positive attitude. Writers are not very often born, but they are very often made.

Practice Exercises

1. Put an advertisement up on the bulletin board of your local library, or in the classifieds of your local community arts newsletter, or in the student center of the nearest college or university in which you request members for a new independent writing workshop group. Try to keep the membership to around six and meet regularly (either monthly or

bimonthly). Establish some organizational plan to the meetings, e.g., you will first discuss micro points (at the level of the sentence and paragraph) and then turn to macro points (overall structure). You may also want to have a rule that the person whose work is being critiqued does not respond until all comments have been made by all members. Make certain that everyone has a copy of the work to be discussed a few days before the meeting, so that an intelligent response can be generated. Make all references in the critique as impersonal and objective as possible. Politeness is the rule of the day. As you receive these comments, bear in mind that the respondent may not be a member of your readership—someone who would buy the book—and that their comments may or may not have relevance to what you are trying to accomplish. Taking well-intentioned advice from the wrong source has destroyed or crippled many a work in workshops. On the other hand, an alert reader who is in your potential constituency can absolutely benefit the work in his or her comments. The important thing is to distinguish between the two.

2. Organize a discussion group on one of the following: a regional issue (such as writers who are concerned with developments in the Greater Yellowstone Ecosystem), a particular animal or plant (such as writers concerns with the spotted owl, or the Florida panther, or the Mexican wolf), or a particular environmental movement (such as deep ecology). You may ask for members to prepare writings specifically on the topic of the conference, and then gather them together in an anthology to be published. Try to coordinate your activities with a local college or university campus (also a good place to find vacant classrooms in between sessions) and utilize the media to promote attendance.

3. Obtain information from a creative writing programs at a school whose faculty, orientation, or geographic location interest you. Ask for the names of graduates and contact them—What did they think of the program? Were they able to find employment or publish books as a result of their studies? Would they recommend it to you?

4. Contact the English Department of a local college or university and ask if you can sit in on one or two workshop classes to acquire some sense of how they are conducted before proceeding either with your own independent workshop or with possible enrollment in an institutionally-affiliated workshop.

Chapter Fifteen

Publication

"Monsieur Flaubert is not a writer."
— from a review in *Le Figaro*
of Gustav Flaubert's 1857 novel *Madame Bovary*

Writing was first invented in Mesopotamia about five thousand years ago in order to maintain business records and chronicle the feats and follies of kings. In its earliest form writing was nothing more than pictures, something like what my son draws when I ask him what he wants for Christmas and he can't describe it with words. From these pictographs evolved hieroglyphics—slightly more sophisticated—and from the hieroglyphics came the first alphabets. There is some debate about where alphabets first arose. The Chinese argue for China and the people of the West insist the revolution occurred in Syria and Palestine. In either case, the event took place around 1500 BC, corresponding with the invention of paper and use of papyrus, and scripts gradually developed in China, Mesopotamia, Egypt and India. The codification of language helped to unify the cultures in all four regions. It is probably not a coincidence that published literature arose just as human societies were becoming more urbanized and hence more socially and legally aware (legal in the sense of social contracts). The need was there to make the products of the mind more durable (especially norms of conduct like the Code of Hammurabi in 1760 BC). Literacy for the masses followed shortly. Anyone who has been to Pompeii, which was preserved *in situ* by a volcanic explosion in the first century AD, has seen that the common people of ancient times could read pretty much as our own—the walls are covered with Roman political slogans, love declarations, and doomsday graffiti.

The invention of the printing press forever changed human life and made possible publishing as it is known today. Prior to that time, every work of literature had to be meticulously copied out by sore-handed scribes. Once the German inventor Gutenberg (1390–1468) gave the concept of movable type and oil-based inks to the world, books could be mass produced. By 1475 the printing press had found its way to England, and a century later—not surprisingly—we have the poetry of Sir Philip Sydney, Christopher Marlowe, William Shakespeare, and Ben Jonson, as well as the prose of Roger Bacon (*Novum Organum*) and Richard Haklyut (*Principall Navigations*), not to mention Michel de Montaigne (*Essais*) across the Channel in France. Technology in this instance had an electrifying effect on literature, sort of like that of Thoreau's spring sun on the fertile earth of Walden Woods. More recently, with the invention of photo offset printing earlier in the century and the use of desktop computers for book production in the 1980s, the whole concept of publishing has taken another giant leap forward. The Macintosh computers produced since 1984 by Apple have been particularly beneficial for writers and students of writing. Even someone as machine-averse as the author of this book can easily operate such a system. It is now possible for a single person, armed only with a user's manual, an Apple Computer, an optical scanner, and a laser printer, to accomplish by pushing a few buttons while talking on the phone before lunch what a whole army of scribes could not have achieved in a decade of thankless toil in the days of Caesar and Christ. All indications are that the next generation of computers will be even faster and more efficient. Still over the horizon, but being worked on daily by researchers, are three dimensional crystals that will be capable of even more impressive miracles of information processing and storage. What all of this says is that writers can look forward to a future in which technology will continue to liberate and empower.

It begins with a dream, as I said so many chapters ago in the Preface. A notion that you can put together words in a way not seen before. That you can create an artifact that will endure as long as the honeybee trapped in the amber of a Jurassic tree. Part of the impulse comes from just that—the desire to somehow outlast your final breath. Another part comes from the wish to entertain and to edify, to share what is best, and worst, from your life with others. So that is where you begin—with an original idea for an article, an essay, a poem, a story, a novel, a nonfiction book. The idea persists. It refuses to die. You begin to make notes. To think. To plan. And then to wonder, if you have never done it before, where do I go from here? The next stage in moving this fertilized egg along towards the birth of publication is the proposal. A proposal is what you transmit to a publisher—the person with the capital to publish and promote your creation—to try to convince him or her that it would be worthwhile to invest in your idea. As you consider your first proposal, never forget what the bottom line is here. Shakespeare knew this (returned to Stratford and bought the biggest beam and plaster mansion in town). Twain was aware of it (made half a million dollars for Grant's destitute widow by publishing the general's memoirs). Hemingway swore by it (always wrote his stories and novels with an eye to the Hollywood film rights his ex-wives loved). Even Edward Abbey was not shy about it ("Never had to work a day in my life after *Desert Solitaire* was reissued," he used to say). All successful writers (in the material sense) have known that there are two ways to work in this world—hard and smart—and have deliberately chosen projects that would be both pleasurable to write and profitable to sell. Those who believe in aestheticism—art for art's sake—certainly have a valid approach, too (and I loyally swore by that admirable creed all those years I starved writing poetry), but this section of the book is for those who may have other equally legitimate interests.

The proposal consists of a cover letter, a brief description of the book, an outline, a list of your publications, a short biography, a preliminary marketing plan for the book, and a sample of the work (generally at least 10 percent). Think like a car salesman or a real estate agent as you prepare the proposal. Remember that you will have to sell the book to the acquisitions editor and then provide her with the tools to sell your book to the editorial board at their monthly or bimonthly meeting. The editorial board are primarily business people—they have light bills to pay and pension plans to fund—and so you can expect them to be more skeptical than the acquisitions editor. Try to give the acquisitions editor as much information as you can—graphics are particularly helpful (cibachrome prints from color slides if there are to be color illustrations), as are past reviews of your books and letters of endorsement from experts in the field. The proposal can also be sent to an agent (I've never used one, but they are helpful for nonfiction projects with national appeal and are almost universally required for fiction). By studying a current edition of *Writer's Market* you can determine which publishers are most suited for your work (for example, a West Coast publisher for a field guide to Pacific Coast birds, a New York publisher for a major collection of essays on Yosemite National Park with broad national appeal).

You basically have three choices as to publishing: a regional press, a university press, or a national publisher. Most of the national publishers are in New York City. The upside with them is that they are generally better capitalized than smaller firms; hence the advances are larger and more resources exist for promotional tours. The downside is that your works may not stay in print as long as with a smaller firm, the editorial staff sometimes turns over rather frequently (editorial continuity is important to an author, particularly a young one in need of guidance), and sometimes the marketing is not as well planned as with a smaller firm whose livelihood depends more on the success of all titles on the list. I have worked primarily with university presses in my career. It used to be that university presses dealt almost exclusively with scholarly texts. During the 1980s, however, a radical transformation occurred as the financial situation for higher education began to deteriorate. No longer could university presses be supported by general funds from the college. They had to enter the marketplace and sell books that would show a profit. As a result, university presses now publish poets, novelists, and nonfiction writers whose works have mass appeal. The University of Chicago Press, for example, published Norman MacLean's novel *A River Runs Through It* and Harvard University Press published several of the trade works of two-time Pulitzer prize winner Edward O. Wilson. The chief advantage with the university presses is that editors tend to stay in place for long blocks of time; hence, long-term working relationships can be built and nurtured. University presses, however, do not usually have the resources for wide-scale promotion, and authors are sometimes dismayed at how good books wither on the vine. Also, because every acquisitions choice must go through an editorial board of professors, decisions to go to contract can take a long time. Most writers will find the editorial staff, the art production staff, and the marketing people to be consistently excellent, in terms of professionalism, with university presses. The last choice is a small or medium-sized regional press, like Alaska Northwest Publishing in Seattle, Washington, or Pruett Publishing in Boulder, Colorado. Here the advantages are much the same as with the university press—long-term staff and a careful dedication to excellence—as are the disadvantages—small capital resources for advances and marketing plans. Many of these publishers produce superb books, and enjoy surprising profits on their ventures. Oftentimes they will sell off the paperback rights to New York firms with the ability to distribute a book more efficiently nation-wide.

A month or so after you submit the proposal a response will arrive from the publisher. Generally, my acceptance rate runs about 20 percent. That is, for every five publishers I contact with respect to a project, one will accept it. Do not be discouraged by rejection letters—everyone gets them. The rejection letter does not mean the editorial staff does not like your writing, only that they do not see it as a potential business investment for their company. There is a significant difference. If you believe in a project, keep trying. California writer Evan Connell sent his fictionalized treatment of the Battle of the Little Bighorn (*Son of the Morning Star*) to eleven east-coast publishers before finding a local one—North Point in San Francisco—that begrudgingly accepted it. The book made the New York Times best-seller list, was sold to the Book of the Month Club for $250,000, and was later adapted for an ABC-TV mini-series. More recently, an Athabaskan woman from Fort Yukon, Alaska, had her native folk story (*Two Old Women*) rejected by so many publishers she finally gathered up $200 each from ten friends and published it herself, via a local publisher in Fairbanks. The book was published to immediate critical acclaim. The entire first edition sold out within two weeks, foreign translation rights were licensed, and movie rights were sold. Every time she went to a book signing there was a line of autograph-seekers stretched outside the door.

Having received a letter of interest from a publisher, you have now reached a point of great interest to many writers: the contract. I generally ask the editors to prepare something reasonable, to treat me as they would wish to be treated. Most editors in my experience can then be trusted to prepare something that entitles both publisher and author to a fair and reasonable profit for their labors and risks. In two cases—*Wildlife in Peril* in 1987 and *The Great Bear* in 1991—I've donated all my royalties to charities, so the contract was really a moot point. A typical contract might grant an author an advance of $5,000 against a royalty of 10 percent on the first 5,000 copies, 12.5 percent on the second 5,000 copies, and 15 percent thereafter. As a rule of thumb, the advance is an approximation of what the first year earnings on the book are expected to be for the author, i.e., the book in the first year will sell 5,000 copies at one dollar per copy for the author, hence an advance of $5,000. The advance is generally paid in two installments: upon signing the contract and upon delivering an acceptable manuscript. Sometimes a small reserve is held back pending the final return of corrected page proofs to the publisher. The book must then make a royalty profit of at least $5,000 before the author receives any further money.

About one or two months after the manuscript is accepted it will come back to you, corrected by the copy editor. The copy-editor is often a freelance person hired on a contract basis by the publisher. Their backgrounds are varied. I have had retired school teachers, former college professors, and professional librarians. My favorite was the first, a former high school English teacher, who worked with me on the three books I published with Oxford University Press and seemed to understand the way I wrote. Copy-editors are experts at grammar and usage. Generally you will also receive a cover letter from this editor, as well as a style sheet, which is a running account of the editing decisions on such issues as spelling and citations. Most of the changes will be confined to mechanics—grammar, spelling, and punctuation. Sometimes there will be substantive changes suggested, marked with query slips, and you need to pay close attention to these. It is important that you defend the integrity of your style, if the changes are diluting, subverting or interfering with the way you express yourself. On the other hand, these recommended substantive changes can sometimes clarify translucent passages and enhance the effectiveness of your arguments or descriptions. For example, I once had an editor recommend parallel chapter titles. This turned out to be a helpful recommendation. I dispensed with the ones previously written and with a little more work achieved much better results; the result was a much more unified book. There will also be instructions for typesetting the text. You should work on this manuscript as quickly and efficiently as possible, answering every query as a matter of course. A speedy and thorough turn-around does two things: it speeds up production of your book and enhances your reputation as a professional writer.

Not long after you return the corrected text you will receive the galleys. The book has now been inputted and a digitized phototypositor machine used to set the type. It is important that you proofread the galleys carefully, because corrections in the page proof stage can be very expensive. The page proofs, a second generation set of proofs in which the text has been set to pages, will also be proofread by the compositor and by the publisher's proofreader. At about this same time you will receive the author's questionnaire from the marketing department. This form provides the in-house advertising specialists with valuable information that will be used to promote the book. Often a request will be made for a recent black and white photograph that can be sent ahead to reviewers and the media. Also during this time, the book, if it is a work of nonfiction, will be indexed, either by the author (I've done eight so far) or by a professional indexer (ordinarily a professional librarian working on a freelance basis).

The next and most complicated stage is printing the book. The printer makes the press plates that carry the images—text and graphics—for the book. The press plates are prepared from the mechanicals (the pages as they have been set up by the book designer). In a process known as photo offset lithography the printer uses the press plates to deposit the image of each page on its respective piece of paper. A printing press consists of a multitude of rotating cylinders through which the paper travels (There are two alternatives to offset printing—letterpress and gravure—and no doubt there will be other technological improvements in the future that will simplify and speed up the process). Once the pages, called signatures, are printed, folded, and sewn together, they are glued, trimmed, backed and jacketed. We now have a book, ready to be boxed and sent to the warehouse, where the distributor will put them in the bookstores for the potential reader.

Months before the book appears in bookstores, the senior staff of the publishing company—the acquisitions editor, the senior editor or publisher, the marketing director, the sales manager, the art director, the managing editor—have conferred and agreed upon a marketing plan for the book. Depending on the circumstances, this can involve anything from doing nothing to preparing an elaborate promotional tour across the country. Most books fall somewhere in between. Typically, the author will be interviewed, either in person or over the phone, by various book reviewers, will make a few appearances at local or regional bookstores, and will perhaps give a reading or two. Several years ago, I was sent on a four-city author's tour in the Pacific Northwest that illustrated to me some of the potential virtues and problems with this form of promotion. The book was published in New York City, and so the publisher hired a regional marketing specialist to plan the tour. Their only contact with this individual had been at the annual American Bookseller's Association meeting the previous May; they had contracted based on little more than a business card and a handshake. As it turned out, the regional person whom they barely knew did an unsatisfactory job in preparing the plan: radio station's in which I was to be interviewed had not received the book in advance, there was virtually no promotion of book store appearances, and there was no use of the newspapers in the cities visited. The last is, in my judgment, the most important tool in promoting a book during an author's tour. Why? Because just about everyone in the group of people likely to buy a book reads the newspaper on a daily basis. Not everyone listens to a particular radio station, or, smaller still, decides to visit a bookstore at the particular hour an author will be there. A well-placed interview with the author in a newspaper can reach a vast audience with good effect. The tour was a very costly undertaking for the publisher, and unfortunately their author had about as much effect on those markets as the shadow of a cloud passing overhead. A word to the wise—become actively involved in any plans to market your book, from the initial author's questionnaire to the final battle plan. Don't hesitate to speak up, as I was, out of politeness. You and your publishers are partners and it is important that you work cooperatively together to enhance profits.

Sometime after the book appears you will begin to receive reviews and letters. Over the years I have been in the position of being both a book reviewer and a book author, and so I can see both sides. My policy on reviews is that if a book is really bad I won't review it. Not because I don't want to be critical—I can be—but rather because I know how much work goes into writing, editing, publishing, and promoting a book and I just can't be a part of demolishing something that so many have worked so hard for. Let the consumer do that, in the marketplace. Generally in my reviews I try to explain what the book is about, to accentuate the positive, and to touch on whatever flaws might exist. Sometimes I use the book as a point of departure for discussing a larger subject. These sorts of reviews are known as essay/reviews and can run up to 2,000 words or more. You find them most frequently in *The New York Review of Books* and in the scholarly journals where contributors have more space than in the newspapers or the Sunday book review sections. Don't be dismayed by bad reviews.

Treat them with humor, if anything, as Edward Abbey did when he wrote in his little book of posthumously-published quotations *A Voice Crying in the Wilderness* that "I've never yet read a review of one of my books that I could not have written much better myself."

Letters can be quite interesting, ranging from lengthy invectives to friendly salutations. Two stand out in my experience. The first was received in response to the hiking guide I wrote for the Gila Wilderness in 1988. The writer, a fire lookout in the Mogollon Mountains, claimed that my book would draw a biblical number of backpackers into the area. I argued back, as John Muir did in his book on the national parks, that we needed more, not fewer, foot soldiers in our little green army. I also stated it was doubtful that hikers could ever impact the area as much as a century of year-round livestock grazing and timber cutting had. Without widespread support, I concluded, these wild areas will be degraded and developed by special interests undeterred by public outcry. Several years later, the essence of this argument was made painfully clear to me as a developer filed a plan to build a 1,790 acre ski resort deep inside the wilderness area. Fortunately, the people were there to rise up in opposition and kill the plan before it moved to the environmental assessment stage. More recently, I received a five page single-spaced letter from an official at Yellowstone National Park complaining about my treatment of the park in a book entitled *Out Among the Wolves*. Basically, the staff at the park, or some of them anyway, wanted me to adhere to the "environmentally correct" history of park management with respect to wolves. They took umbrage at my stating that the eradication of wolves in Yellowstone from 1915–1915 was the result of the most egregious wildlife management decision ever made by the federal government. They also didn't care for my treatment of the same agency on their slaughter of grizzly bears in the early 1970s following the dump closings. I replied that the independent writer has a responsibility to get at the truth and take an informed position, regardless of the popularity of the position. Our readers expect nothing less than this. These two examples illustrate what I consider to be a very important part of the book publishing process—to follow up on important issues that arise from publication and to defend your book if it is unfairly attacked or misrepresented. If you receive long letters that readers have obviously worked on, I believe you have a duty to respond in similar length to their concerns.

There is nothing like that first published article or book. I will never forget either. My first article was a short regional piece on trout fishing in the La Garita Wilderness of southern Colorado that appeared in *Outdoor Life* about twenty years ago. When the magazine arrived, and the check for $60, I must have re-read that thing a dozen times, and kept the issue proudly by my old Royal portable typewriter as I began to write successors for the same magazine. Pitiful as the little piece is by most standards, it validated my choice of career and gave me hope for the future. Similarly, my first book—a guide to what was then the most heavily used wilderness area in the U.S. (the Indian Peaks west of Denver)—occupies a special place on the shelf. I hiked every one of those high mountain trails, took hundreds of pictures, composed the maps, and even slipped in a short essay at the end, so that I could call myself an essayist, which, it seemed to me, was something slightly more than the journalist I had been. Many books have followed, but I've never surpassed the excitement of the day I received that glossy green volume with the photograph of Brainard Lake (most beautiful lake in the world) on the cover. It is my hope in writing this book, and sharing it now with you, that someday I will receive a letter from a reader saying that it inspired them to their first publication. As you move toward that moment, bear in mind a sentence from Thoreau's journals:

> I should like to keep some book of natural history always by me as a sort of elixir, the reading of which would restore the tone of my system and secure me true and cheerful views of life.

Aim to create a work that he would have wanted beside him.

Practice Exercises

1. If you live near a medium-sized or large city, you probably have access to a regional publisher or a university press. Make arrangements to visit the offices and interview key personnel, such as the marketing director, the acquisitions editor, the managing editor, the art director, the senior editor or publisher, and others. Study a recent catalog and ask questions about the business of making books. You might also inquire as to how personnel entered the industry, with an eye as to how you could do the same. Over the years, many writers, from T.S. Eliot to David Brower, have worked in publishing as editors; it is a natural environment for people who love books.

2. Similarly, if you live near a metropolitan area, you can tour the manufacturing facilities of a book printing company and thus gain helpful insights into the final stages of book production.

3. While visiting the publisher in exercise one, obtain a copy of a publication contract and carefully study it, paying close attention to the sections outlining the responsibilities of the two parties, the nature of the royalty structure, and the various rights assigned to the publisher and author.

4. Contact an agent (or the chief association—Society of Authors' Representatives, 10 Astor Place, 3rd Floor, New York, New York 10003) and obtain a standard contract. As with the publishing contract, slowly read through it, familiarizing yourself with the primary features, and the clear advantages and disadvantages of using an agent based on the arrangements stipulated. What is the extent of the agent's representation? What is the agent's commission? Who pays for phone calls, postage, and so forth? How is the relationship terminated?

5. Using the contract, role-play with a fellow writer, with one person assuming the role of the publisher and the other person assuming the role of the author. Negotiate over various parts of the contract in an effort to understand how each side has legitimate concerns over being properly compensated for work and risks undertaken.

Chapter 16

Nature Writing and Environmental Activism

It is always a writer's duty to make the world better.
— Samuel Johnson, Preface to *Shakespeare*

Since the beginning, nature writing has been an active force of social change, engaging controversial issues of the day and stimulating public discourse on topics of concern. The history of nature writing is, to no small degree, the history of the environmental preservation movement in the United States. Although literary commentary upon nature first began in classical times—even Plato and, later, Pliny commented upon the deforestation of the Mediterranean—it was not until the modern era that the effects of technology and population growth were such that professional writers began to use literature as a political instrument *vis-à-vis* conservation.

One of the first influential documents to appear in this respect was George Catlin's 1841 account of travel in the Far West, in which he made a plea for "a nation's Park, containing man and beast, in all the wild and freshness of their nature's beauty." Catlin was a far-sighted visionary. A quarter of a century before the railroads brought a flood of homesteaders to the high prairie, the artist-writer was already predicting the extinction of the bison and the pronounced effects this would have upon both the natural landscape and Native American culture. Fifty years after his book was published, in March of 1872, President Ulysses S. Grant made Catlin's youthful dream a reality when he signed into law the legislation creating Yellowstone National Park—the world's first national park—in what was then the Wyoming Territory.

Not long after, early activist-writers such as John Muir and Theodore Roosevelt began to write magazine articles and publish books advocating the formation of new national parks, the conservation of endangered species (such as bison), and the more scientific management of the national forests (both wrote eloquently of the need). Roosevelt later put his words into action when he became President, setting aside over 150 million acres as national forests and/or national monuments (including the Grand Canyon). Similarly, he explored both East Africa and Brazil after he left the presidency, and argued for nature conservation in other parts of the world:

> All civilized governments are now realizing that it is their duty here and there to preserve, unharmed, tracts of wild nature, with thereon the wild things the destruction of which means the destruction of half the charm of wild nature . . . There should be certain sanctuaries and nurseries where game can live and breed absolutely unmolested [from Wild Africa].

In his calls for the world-wide preservation of nature, Roosevelt anticipated the work that organizations such as the World Wildlife Fund and the Nature Conservancy would later undertake in the twentieth and twenty-first centuries.

Like Theodore Roosevelt, John Muir had a quiet but unshakeable belief that words must be wed to action. It is difficult to imagine the circumspect Ralph Waldo Emerson or the owlish John Burroughs using blunt language like this:

> These temple destroyers, devotees of raging commercialism, seem to have a perfect contempt for Nature, and instead of lifting their eyes to the God of the Mountains, lift them to the Almighty Dollar.

Muir took nature-writing out of the airless parlor rooms and polite lecture halls and made it something to fight for in the federal courthouses and state capitol committee rooms. As a result of his firm but rational militancy, he was instrumental in forming Yosemite National Park, Sequoia National Park, and General Grant National Park. Although he lost the battle to defeat Hetch-Hetchy Dam (which Roosevelt supported as a water source for San Francisco), his heroic David-and-Goliath effort inspired later dam-fighters (such as David Brower and Ansel Adams). Today Muir's legacy lives on—in the Sierra Club which he founded, in the John Muir Trail of the High Sierras, in a national monument north of San Francisco protecting a rare redwood grove, and in the half-million acre John Muir Wilderness. All of this arose from Muir's devotion, expressed in a dozen books and hundreds of articles, in the spiritual reclamation of the West.

Soon these two early pioneers were joined by other kindred spirits, including such luminaries as John Van Dyke and Mary Austin, both of whom wrote eloquently of the need to appreciate and preserve the desert environment. During the 1920s an entirely new generation of writers—most notably Aldo Leopold and Bob Marshall—proposed that the nation establish permanent wilderness areas in the national forests of the west. It was Leopold who wrote, in an oft-quoted passage from his essay "The Green Lagoon":

> Man always kills the thing he loves, and so we pioneers have killed our wilderness. Some say we had to. Be that as it may, I am glad I shall never be young without wild country to be young in. Of what avail are forty freedoms without a blank spot on the map?

As a result of Leopold's indefatigable efforts the world's first wilderness area, the Gila, was established in 1924 in southwestern New Mexico. As it was originally conceived, the Gila was nearly the size of Yellowstone, and protected over a thousand square miles of rugged volcanic mountains and canyons in the former homeland of the Chiricahua Apache. The Gila Wilderness Area was soon followed by many others, in Colorado (Upper Rio Grande and Flattops) and elsewhere in the West.

The writings and advocacy of Bob Marshall were particularly important in the formation of these later wilderness sanctuaries. Marshall, who worked as a senior official for the Forest Service (director of the division of forestry and grazing), labored tirelessly on behalf of the first primitive and wilderness areas. His spirit was uncompromising: "We want no straddlers, for in the past they have surrendered too much good wilderness and primeval which should never have been lost." In his short life, Marshall went on to explore the *terra incognita* of northern Alaska—his book *Alaska Wilderness* became an instant classic—and to form, with his sizeable inheritance, the influential Wilderness Society. His manifold contributions were later acknowledged by a grateful posterity in the formation of the 950,000-acre Bob Marshall Wilderness Area in the northern Rockies of Montana.

During the 1940s and 1950s attention turned more from forming new areas to preserving what had already been set aside from the effects of increasingly intrusive highways, commercial development and dam-reservoir projects. The most effective writer-advocates in this period included William O. Douglas, who served as a Supreme Court Justice, Wallace Stegner, who taught literature and writing at Stanford University, and Ansel Adams, who used his images and his writings to oppose all who would diminish the landscape. Sierra Club executive director David Brower, working with such advocates as Stegner and Adams, successfully defeated several dam projects, including a notorious proposal for a dam on the Green River in Dinosaur National Monument. Adams, who also served on the Board of Directors of the Sierra Club, often gave lyric expression to the beliefs at the heart of the environmental movement in his surprisingly potent writings:

> In contemplation of the eternal incarnations of the spirit which vibrate in every mountain, leaf and particle of earth, in every cloud, stone and flash of sunlight, we make new discoveries on the planes of ethical and humane discernment, approaching the new society at last, proportionate to nature.

One of the most resonant voices for nature during this period was Rachel Carson, whose numerous books about the sea elevated consciousness and helped lead to the formation of national seashores. Her bestseller *Silent Spring*, which documented the pernicious effects of DDT, so inspired President Kennedy that he formed a presidential commission that led to the banning of the chemical in the United States. Another powerful writer who appeared on the scene in the 1960s was Edward Abbey. Like Muir and Roosevelt, Abbey believed that words were useless unless they were married to action. His life was dedicated to the proposition that one person can, and should, make a difference. Like his predecessors, Abbey was a foe of all who would diminish the landscapes of the American West. He used language just as militant (and, in a way, Jeffersonian) in his defense of nature:

> If the wilderness is our true home, and if it threatened with invasion, pillage and destruction—as it certainly is—then we have the right to defend that home, as we would our private quarters, by whatever means are necessary . . . not to defend that which we love would be dishonorable (from the essay "Eco-Defense").

The 1980s and 1990s saw the rise of three primary writer-advocates: George Schaller, Peter Matthiessen and E. O. Wilson, each of whom worked independently around the globe on nature's behalf. Schaller traveled extensively in southern and eastern Asia, studying animals as diverse as the snow leopard of Nepal and the wild antelope herds of Outer Mongolia. His many formidable works from this part of the world include *The Deer and the Tiger* and *Mountain Monarchs*. Peter Matthiessen became devoted to the preservation of the Asian cranes, as well as the Siberian Tiger (*Tigers in the Snow*) and the endangered fauna of central and southern Africa (*African Silences*). Edward O. Wilson, who has twice been awarded the Pulitzer Prize for general nonfiction, has recently written such popular works as *The Diversity of Life* and *The Future of Life*. He has become a champion of biodiversity, and his writings raise difficult questions for human civilization as it enters the third millennium.

Other writers work closer to home. Terry Tempest Williams, writing from Moab, Utah, has become a staunch defender of the high desert, which is one of the most fragile environments in the west. Her most recent book, which was entitled simply *Red*, gathered together her various articles, essays and bits of congressional testimony on the arid landscapes of southern Utah. Further north, Montana writer Annick Smith has assembled an anthology of essays, fiction, and poetry entitled *Headwaters*, in an effort to save the endangered Blackfoot River from the devastating effects of a proposed gold mine. The same river served as a central symbol in Norman MacLean's 1976 novella *A River Runs Through It*, which was later made into a film directed by Robert Redford and starring a young Brad Pitt. Forty-nine authors are featured in the book, including Rick Bass, James Welch, Peter Fromm, Jim Harrison, David James Duncan, and William Kittredge.

My own efforts have been most concerned with the endangered species, especially predators, of the American West. In the mid-1980s, when I first advocated the restoration of the lynx, wolverine, wolf, and grizzly to Colorado in *Wildlife in Peril*, I was regarded, at best, as a dreamer. I can still recall appearing on radio and television programs and basically being laughed at by the audience and the commentator for an hour. I was a good sport, though, because I knew that there is nothing as inevitable as a progressive idea founded in common sense. I understood that all that was required on my part was patience and dedication. Today, nearly twenty years later, wolves have been restored to Wyoming and Idaho, lynx have been restored to Colorado, and several Colorado wolverine studies have been completed. Additionally, a private group has now submitted a detailed plan to the federal government for the restoration of the grizzly to the mountains of southern Colorado.

What all of this means is that a work of nature writing can perform a number of functions, ranging from the purely aesthetic—celebrating the beauty of the natural world—to the strategic and political—moving a city or a state or a nation state along a particular compass heading toward progress. Each writer must make a decision at some point as to what extent this consideration will influence his or her writing. In some cases, the practical, goal-directed mode of thinking will predominate. The writer will devote years, even decades, toward one or several conservation objectives, and his or her writing will clearly serve these ends. In other cases, the writers may choose a less direct approach, in the belief that anything that heightens awareness will inevitably advance larger causes. Either choice is valid, and all literature no doubt helps human progress in ways small and large. All that matters in the end, really, is the integrity, craft and vision of the writing. That and the fact that the writer was down there on the playing field striving to defend the earth.

Practice Exercises

1. Create an anthology on a timely issue of import in your home region. Aim for a melody of voices that powerfully supports a single theme. Specific topics might include the following: the status of the eastern gray wolf in New England (i.e., supporting restoration), the challenges facing the Hudson River Valley, the plight of the Everglades ecosystem, the state of the grasslands, the euphemistic regression known as the public land user fee, the need for a comprehensive recovery plan for the southwestern jaguar (Arizona, New Mexico and Texas). If the anthology includes a diverse collection of high quality writings (normally at least eleven or twelve authors, ranging from the iconic to the emerging), the editor should easily find a university press or regional publisher enthused about publishing the book.

2. Organize a public reading of writers in your area on an issue of mutual concern raise consciousness. The event might include an exhibit of landscape or wildlife photographs or paintings, or even music, in celebration of some aspect of nature. Specific topics could be drawn from the list above, or inspired by some particular cause unique to your home area.

3. Write a nature essay, or book, that is devoted to a specific cause or goal—the restoration of the Condor to the Santa Lucia Mountains of central California, the dwindling of salmon and sea otter populations in certain parts of Alaska, the clear-cutting of southern old growth forests, the issuing of oil and gas exploration permits in sensitive desert regions, the threat of mineral development in the Arctic National Wildlife Refuge. After publication, work in a substantive way to marry

those words to a particular action that advances the cause (this could range from something as quiet as a petition drive to testimony at a congressional hearing to the organization of a public rally).

Appendix

Recommended Readings

The Norton Book of Nature Writing, edited by Robert Finch and John
Elder (Norton, 1990)
This Incomparable Lande: A Book of American Nature Writing,
edited and with an introduction by Thomas J. Lyon (Houghton
Mifflin, 1989)
Writing Natural History: Dialogues with Authors, edited by Edward
Lueders (University of Utah, 1990)
On the Teaching of Creative Writing, by Wallace Stegner (University
Press of New England, 1988)
Teaching Environmental Literature: Materials, Methods, Resources,
edited by Frederick O. Waage (The Modern Language Association of
America, 1985)
The Wilderness Reader, edited by Frank Bergon (Mentor, 1980)
Western American Quarterly (quarterly, published by the Department of
English, Utah State University, Logan, Utah)
Eternal Quest: The Story of the Great Naturalists, by Alexander Adams
(Putnam, 1969)
Speaking for Nature, by Paul Brooks (Sierra Club, 1983)
*Traces on the Rhodian Shore: Nature and Culture in Western Thought
from Ancient Times to the End of the Eighteenth Century*, by
Clarence Glacken (University of California 1967, reissued 1992)
*The Lay of the Land: Metaphor as Experience and History in American
Life and Letters*, by Annette Kolodny (University of North
Carolina, 1975)
Wilderness and the American Mind, by Roderick Nash (Yale University
Press, 1967, revised 1982)

Creative Writing Programs

Creative Writing Program
Department of English
University of Montana
Missoula, Montana 59812
406-243-5231

Creative Writing Program
Department of English
University of Arizona
Tucson, Arizona 85721
602-621-3880

Creative Writing Program
Department of English
University of California, Davis
Davis, California 95616
916-752-2281

Creative Writing Program
Department of English
University of California, Irvine
Irvine, California 92717
714-856-6712

Creative Writing Program
Department of English
Iowa State University
203 Ross Hall
Ames, Iowa 50011
515-294-2180

Creative Writing Program
Department of English
University of Denver
Denver, Colorado 80208
303-871-4387

Creative Writing Program
Department of English
University of Hawaii
1733 Donaggho Road
Honolulu, Hawaii 96822
808-948-7619

Creative Writing Program
Department of English
Florida State University
Tallahassee, Florida 32306
904-644-4230

Creative Writing Program
Department of English
University of Alaska
Fairbanks, Alaska 99775
907-474-7193

Creative Writing Program
Department of English
Indiana University
Bloomington, Indiana 47405
812-855-8224

Creative Writing Program
Department of English
Cornell University
Goldwin Smith Hall
Ithaca, New York 14853
607-255-6802

Creative Writing Program
Department of English
University of Utah
Salt Lake City, Utah 84112
801-750-2733

Creative Writing Program
Department of English
Bowling Green State University
Bowling Green, Ohio 43403
419-372-8370

Creative Writing Program
Department of English
The University of Kansas
Lawrence, Kansas 66045
913-864-4520

Creative Writing Program
Department of English
University of Washington
Seattle, Washington 98195
206-543-2690

Creative Writing Program
Department of English
University of North Carolina
Greensboro, North Carolina 27412
919-334-5459

Creative Writing Program
Department of English
State University of New York
Buffalo, New York 14260
716-636-2575

Creative Writing Program
(confers no graduate degree,
but offers workshop)
Department of English
Stanford University
Palo Alto, California 94305
415-723-2637

Creative Writing Program
Department of English
Louisiana State University
Baton Rouge, Louisiana 70803
504-388-2236

Creative Writing Program
Department of English
University of Nebraska
202 Andrews Hall
Lincoln, Nebraska 68588
402-472-3191

Environmental Organizations

The Wilderness Society: 900 17th St. N.W.
Washington, D.C. 20006
202-833-2300

World Wildlife Fund: 1250 24th St. N.W.
Washington, D.C. 20037
202-293-4800

The Nature Conservancy: 1815 N. Lynn St.
Arlington, VA 22209
703-841-5300

National Audubon Society: 700 Broadway
N.Y., N.Y. 10003
212-979-3000

Sierra Club: 730 Polk St.
San Francisco, CA 94108
415-776-2211

Greenpeace USA: 1436 U St. N.W.
Washington, D.C. 20009
202-462-1177

Environmental Defense Fund: 257 Park Ave. S.
N.Y., N.Y. 10010
212-505-2100

National Wildlife Federation: 1400 16th St. N.W.
Washington, D.C. 20036
202-797-6800

Defenders of Wildlife: 1101 14th St. N.W.
Washington, D.C. 20005
202-682-9400

Nature Resources Defense Council: 40th W. 20th St.
N.Y., N.Y. 10011
212-727-2700

Sierra Club Legal Defense Fund: 180 Montgomery St., Suite 1400
San Francisco, Ca 94104
415-627-6700

Conservation International: 1015 18th St. N.W.
Washington, D.C. 20036
202-429-5660

Writing Conferences

Antioch Writers' Workshop (July)
135 North Walnut Street
Yellow Springs, Ohio 45387
513-767-9112

Bay Area Writers' Workshop at Mills College (August)
PO Box 620327
Woodside, CA 94602
415-430-3127

Bennington Writing Workshops (July)
Bennington College
Bennington, Vermont 05201
802-442-5401, ext 367

Charleston Writers' Conference (spring)
College of Charleston
Charleston, South Carolina 29424
803-792-5664

The Deep South Writers' Conference (September)
USL Box 44691
University of Southwestern Louisiana
Lafayette, LA 70504-4691
318-231-6908

Desert Writers' Workshop (October)
Canyonlands Field Institute
PO Box 68
Moab, Utah 84532
801-259-7750

The Eastern Writers' Conference (June)
Salem State College
Salem, Massachusetts 01970
508-741-6270

Florida State Writers' Conference (May)
PO Box 9844
Fort Lauderdale, Florida 33310
800-351-9278

Hofstra University Summer Writers' Conference (Summer)
U.C.C.E.
Memorial Hall, 232
Hempstead, New York 11550
516-560-5016

Iowa Summer Writing Program (June/July)
Division of Continuing Education
116 International Center
The University of Iowa
Iowa City, Iowa 52242
319-335-2534

Key West Literary Seminar (January)
PO Box 391
Sugarloaf Shores, Florida 33044
305-745-3640

The MacDowell Colony (year-round)
100 High Street
Petersborough, New Hampshire 03458
603-924-3886 or 212-966-4860

The Maine Professional Writers' Workshop (summer/fall)
Rockport, Maine 04856
207-236-8581

Midwest Writers' Conference (October)
Kent State University Stark Campus
6000 Frank Avenue, Northwest
Canton, Ohio 44720
216-499-9600

Napa Valley Writers' Conference (July/August)
Office of Community Education
Napa Valley College
Napa, California 94558
707-253-3070

New York State Summer Writers' Institute (summer)
Skidmore College
Saratoga Springs, New York 12866
919-967-9540

Oklahoma Arts Institute (summer/fall)
720 NW 50th, PO Box 18154
Oklahoma City, Oklahoma 73154
405-842-0890

Santa Fe Writers' Conference (August)
Recursos de Santa Fe
826 Camino de Monte Rey
Santa Fe, New Mexico 87501
505-982-9301

Southern California Writers' Conference, San Diego (January)
3745 Mt. Augustus Avenue
San Diego, California 92111
619-277-7302

Steamboat Springs Writers' Conference (August)
Steamboat Springs Council of the Arts and Humanities
PO Box 1913
Steamboat Springs, Colorado 80477
303-879-4434

The Wesleyan Writers' Conference (summer)
Wesleyan University
Middletown, Connecticut 06457
203-347-9411 ext 2448 or 2547

Yellow Bay Writers' Workshop (summer)
Center for Continuing Education
University of Montana
Missoula, Montana 59812
406-243-6486

References Cited

Abbey, Edward. *Abbey's Road*. New York: Dutton, 1979.

———. *Desert Solitaire*. New York: McGraw-Hill, 1968; Tucson: University of Arizona Press, 1990.

———. *Down the River*. New York: Dutton, 1982.

———. *The Journey Home: Some Words in Defense of the American West*. New York: Dutton, 1977.

———. *One Life at a Time, Please*. New York: Henry Holt, 1987.

———. *Selected Essays*. San Francisco: Sierra Club, 1988.

Bartram, William. *Travels Through North and South Carolina, Georgia, East and West Florida, the Cherokee Country, the Extensive Territories of the Muscuogulges, or Creek Confederacy, and the Country of the Choctaws*. Philadelphia: James & Johnson, 1791; Layton, Utah: Peregrine Smith, 1980.

Bass, Rick. *The Deer Pasture*. Fort Worth: Texas A & M University Press, 1985; W.W. Norton, 1988.

———. *Wild to the Heart*. New York: W. W. Norton, 1990.

———. *Winter*. New York: Houghton Mifflin, 1991.

Berry, Wendell. *A Continuous Harmony: Essays Cultural and Agricultural*. New York: Harcourt Brace Jovanovich, 1972.

Beston, Henry. *The Outermost House*. Garden City, New York: Doubleday, 1928; Ballantine, 1971.

Bowden, Charles. *Blue Desert*. Tucson: University of Arizona Press, 1986.

Carson, Rachel. *The Sea Around Us*. Boston: Houghton Mifflin, 1955; Oxford University Press, 1961.

Dillard, Annie. *Pilgrim at Tinker Creek*. New York: Harper & Row, 1974.

Dinesen, Isak. *Out of Africa*. New York: Viking, 1938.

Eckstein, Gustav. *Everyday Miracle*. New York: Harper & Row, 1965.

Ehrlich, Gretel. *The Solace of Open Spaces*. New York: Viking, 1985.

Eiseley, Loren. *The Immense Journey*. New York: Random House, 1957.

Emerson, Ralph Waldo. *Nature*. Boston: Munroe, 1836; Beacon Press, 1986. [This important essay is also available in virtually any anthology of American literature used for teaching post-secondary survey courses].

Fowles, John. *The Tree*. San Francisco: North Point Press, 1986.

Haines, John. *Living off the Country: Essays on Poetry and Place*. Ann Arbor: University of Michigan Press, 1981.

Haines, John. *The Stars, the Snow, the Fire*. Minneapolis: Graywolf Press, 1988.

Harrison, Jim. *Just Before Dark*. Clarke City, Montana: Clarke City Press, 1990.

Hasselstrom, Linda. *Land Circle*. Golden, Colorado: Fulcrum, 1988.

Hemingway, Ernest. *The Green Hills of Africa*. New York: Scribner's, 1935.

Hoagland, Ed. *Red Wolves and Black Bears*. New York: Random House, 1976.

Hubbell, Sue. *A Country Year: Living the Questions*. New York: Harper and Row, 1987.

Krutch, Joseph Wood. *The Desert Year*. New York: Sloane, 1952; Viking, 1963.

Leopold, Aldo. *Round River*. Edited by Luna Leopold. New York: Oxford University Press, 1953.

———. *A Sand County Almanac*. New York: Oxford University Press, 1949.

Lopez, Barry. *Arctic Dreams*. New York: Scribner's, 1986.

———. *Crossing Open Ground*. New York: Scribner's, 1988.

———. *Of Wolves and Men*. New York: Scribner's, 1978.

Lueders, Edward. *Writers on Nature Writing*. Salt Lake City: University of Utah Press, 1989.

MacLean, Norman. *A River Runs Through It*. Chicago: University of Chicago Press, 1976.

———. *Young Men and Fire*. Chicago: University of Chicago Press, 1991.

McPhee, John. *Coming into the Country*. New York: Farrar, Straus & Giroux, 1977, Bantam, 1979.

Marshall, Robert. *Alaska Wilderness*. Berkeley: University of California Press, 1956, 1970.

Matthiessen, Peter. *Sand Rivers*. New York: Viking, 1981.

————. *The Snow Leopard*. New York: Viking, 1978.

Momaday, N. Scott. *The Way to Rainy Mountain*. Albuquerque: University of New Mexico Press, 1976.

Muir, John. *A Thousand-Mile Walk to the Gulf*. Boston: Houghton Mifflin, 1916; Sierra Club, 1990.

Murie, Adolph. *The Grizzlies of Mount McKinley*. Seattle: University of Washington Press 1981.

————. *A Naturalist in Alaska*. New York: Devin-Adair, 1961; University of Arizona Press, 1990.

Murie, Margaret. *Two in the Far North*. New York: Knopf, 1962.

Nabhan, Gary. *The Desert Smells Like Rain: A Naturalist in Papago Indian Country*. San Francisco: North Point, 1982.

————. *Gathering in the Desert*. Tucson, University of Arizona Press, 1985.

Nelson, Richard. *Make Prayers to the Raven: A Koyukon View of the Northern Forest*. Chicago: University of Chicago Press, 1983.

————. *The Island Within*. San Francisco: North Point Press, 1990.

O'Brien, Dan. *The Rites of Autumn*. New York: Houghton Mifflin, 1988.

Olson, Sigurd F. *Runes of the North*. New York: Knopf, 1963.

————. *The Singing Wilderness*. New York: Knopf, 1956.

Pruitt, William O. *Animals of the North*. New York: Harper & Row, 1967.

Pyle, Robert Michael. *Wintergreen: Listening to the Land's Heart*. New York: Scribner's, 1986.

Richardson, Robert. *Henry David Thoreau: A Life of the Mind*. Berkeley: University of California Press, 1986.

Roberts, David. *The Mountain of My Fear*. New York: Vanguard, 1968.

Roosevelt, Theodore. *Hunting Trips of a Ranchman: Hunting Trips on the Prairie and in the Mountains*. New York: Putnam's, 1885.

————. *The Wilderness Hunter*. New York: Putnam's, 1883.

————. *African Game Trails*. New York: Scribner's, 1909.

Schaller, George. *Stones of Silence: Journeys in the Himalaya*. New York: Viking, 1980.

————. *The Year of the Gorilla*. Chicago: University of Chicago Press, 1964.

Seton, Ernest Thompson. *Lives of Game Animals*. 4 vols. Garden City, New York: Doubleday, 1929.

Snyder, Gary. *Earth House Hold*. New York: New Directions, 1969.

Stanley, Henry. *Through Darkest Africa*. New York: Putnam's, 1890.

Stegner, Wallace. *The Sound of Mountain Water*. Garden City, New York: Doubleday, 1969; Lincoln: University of Nebraska Press, 1985.

Steinbeck, John. *The Log from the Sea of Cortez*. New York: Viking, 1951, 1962.

Teale, Edwin Way. *Journey into Summer*. New York: Dodd, Mead, 1960.

Thoreau, Henry David. *Journals*. Vols 7–20 of *The Writings of Henry David Thoreau*. Boston: Houghton Mifflin, 1906; Layton Utah: Gibbs M. Smith, 1984; Princeton: Princeton University Press, 1981.

———. *Walden*. Boston: Ticknor and Fields, 1854; Princeton: Princeton University Press, 1971.

Twain, Mark. *Roughing It*. New York: Putnam's, 1876.

Wallace, David Rains. *The Dark Range: A Naturalist's Nigh Notebook*. San Francisco: Sierra Club, 1978.

———. *Idle Weeds: The Life of a Sandstone Ridge*. San Francisco: Sierra Club, 1980.

———. *The Klamath Knot: Explorations of Myth and Evolution*. San Francisco: Sierra Club, 1983.

———. *The Untamed Garden and Other Personal Essays*. Columbus: Ohio State University Press, 1986.

Williams, Terry Tempest. *Refuge*. New York: Pantheon, 1991.

Wilson, Edward O. *Biophilia*. Cambridge: Harvard University Press, 1984.

———. *The Naturalist*. San Francisco: The Island Press, 1994.

Zwinger, Ann. *Beyond the Aspen Grove*. New York: Harper & Row, 1970, 1981.

Index